AF575292

Reservoir Engineering and Conformal Mapping of Oil and Gas Fields

Reservoir Engineering and Conformal Mapping of Oil and Gas Fields

BY WILLIAM HURST

THE PETROLEUM PUBLISHING COMPANY
TULSA

1421 South Sheridan Road/P.O. Bof 1260
Tulsa, Oklahoma 74101

Library of Congress Catalog Card Number: 78–70551
International Standard Book Number: 0–87814–095–6
Printed in the United States of America

1 2 3 4 5 83 82 81 80 79

THIS BOOK
IS DEDICATED TO MY DEAR DECEASED WIFE
Ruth Prescott Hurst

CONTENTS

PREFACE

THE CHAPTERS IN THIS TEXT are not presented in the chronological order of their development, but are based on interrelated ideas that occurred in the last few years. Thus each chapter in this book is a counterpart of the others.

The first three chapters were presented as a series of lectures in a continuing education held in New Orleans, Louisiana during 1976-1977. The last chapter is a recent development.

The reception received from the audiences in these lectures suggested this information should have broader exposure, which is the reason this text was written and made available to the professional engineer in the oil business.

The first chapter covers Conformal Mapping of Oil and Gas Fields, for which the name of the book is taken.

Conformal mapping for those unfamiliar with the subject is a procedure whereby one surface can be extended to another. In our case, this is the extension of the actual configuration of an oil or gas field to conform to a rectangle. By locating wells in this rectangle, all mathematics that apply to fluid flow can be performed. When this is done the relationship of pressure contours or streamlines can be referred back to the field proper by point-to-point transfer. What is accomplished is reservoir simulation.

The example illustrated in the text is a circular field conformed into a rectangle. The reason for this is we know all the solutions for a well performing within a circle whether for steady state or transient fluid flow. Thus, the choice for a circle permits a check on conformal mapping, amply illustrated in the text.

The next two chapters, Solution of Nonlinear Equations for Flow of Fluids and The Subsidiary Equation, deal with the actual mechanics of fluid flow that pertain to conformal mapping.

The first of these is for the nonlinear flow of gases in a gas reservoir. The gas formula as applies to the diffusivity equation in a reservoir, namely, the balance of gas on an infinitesimal element of sand in the formation, is a nonlinear partial differential equation. The treatment of this equation as applied to the reservoir, in setting up increments of pressure change within the formation and linealizing this formula by holding the changing gas compressibility constant, is shown to give the excellent agreement with results obtained by Bruce et al.

The same idea of linealization has been applied to the Subsidiary Equation that deals with the two-phase solution of oil and its associated gas both flowing to a well. The first is the development of the Subsidiary Equation from the actual equations of the diffusivity formulas for two-phase fluid flow, to give oil saturation in situ with reservoir pressure decline. An illustrative example shows the voidage in an oil field flowing oil and gas to a producing well. Excellent checks between the oil and gas voided within the reservoir and that produced by the well are shown.

The degree of success for this accomplishment is due to the use of the Ei-function of Lord Kelvin, which represents transient fluid flow. This, in the writer's publications, is a departure from the classical Bessel functions used in the past. It is a shorthand notation more readily available to the practical engineer than the more involved methods. This text illustrates its derivation and its solution as presented by other authors.

What will now be mentioned to enlighten the reader on reservoir simulation is that in the rectangle every point in the continuum is influenced by interference from every well in the field. This is carried out by the $P(t_D)$ function covered in Chapter 1, which in turn is the Ei function. Every well for the point under consideration contributes its $P(t_D)$ value. The sum of all these $P(t_D)$ values yields transient fluid flow for which permeability, porosity, sand thickness, etc. apply for the point under consideration.

How the reader wishes to carry out his calculations and procedure — by desk calculator or computers — is his option. With the summation of the $P(t_D)$ values established for transient flow, the reader can simulate a gas or oil reservoir.

The final chapter, In-Fill Drilling, is a practical illustration of reservoir simulation. What is defined here is the delineation of streamlines showing oil and gas flow to producing wells within a reservoir. The application of this work is to show where in-fill drilling can reach oil and gas that have never entered into the mainstream of fluid flow in a field. It has an additional attribute that it presents a blueprint of how a field has performed in the past, and where waterflood or enhanced oil recovery can be instigated to increase the recovery of a field.

The purpose of this text is to make it self-contained for anyone using it. An Appendix is added for anyone interested in the derivation of formulas used in this work; he has to refer to no other source. Likewise, tables have been entered for classroom exercise, if such is considered. This is all the grace the writer can say to the reader for the time being.

WILLIAM HURST

Houston, Texas, U.S.A.
1978

Reservoir Engineering and Conformal Mapping of Oil and Gas Fields

1

CONFORMAL MAPPING OF OIL AND GAS FIELDS

THE MOTIVATION for this chapter is the frustration encountered every time we face a geologist's interpretation of a structural fault block or an oil field. These consist of distorted and contorted patterns representing the reservoir domain.

We often ask, "How do we determine pressure distributions and flow patterns in a typical fault block as shown in Fig. 1-1?"

The proximity of a single well to its enclosures is bound to influence the flow of fluids and alter pressure drops. In the same way, depletion of the reservoir as a whole is influenced by all the enclosures defined by the geologist.

This problem is further compounded by faults and shale lines forming a non-continuous reservoir—a subject to be discussed in the later portions of the chapter.

CONFORMAL MAPPING

The literature and texts give quite a comprehensive discussion on conformal mapping, but very little has direct application to the current problem.

An earlier objective in transferring one domain to the plane of another was the solution of elasticity problems, potential theory, and hydraulics of fluid flow represented by the equation

$$\nabla^2 p = \frac{\partial p}{\partial t_D} \qquad (1\text{-}1)$$

where the right side is zero, corresponding to steady-state fluid flow. Our problem is more involved as we shall see, in that we deal with the unsteady-state flow of fluids.

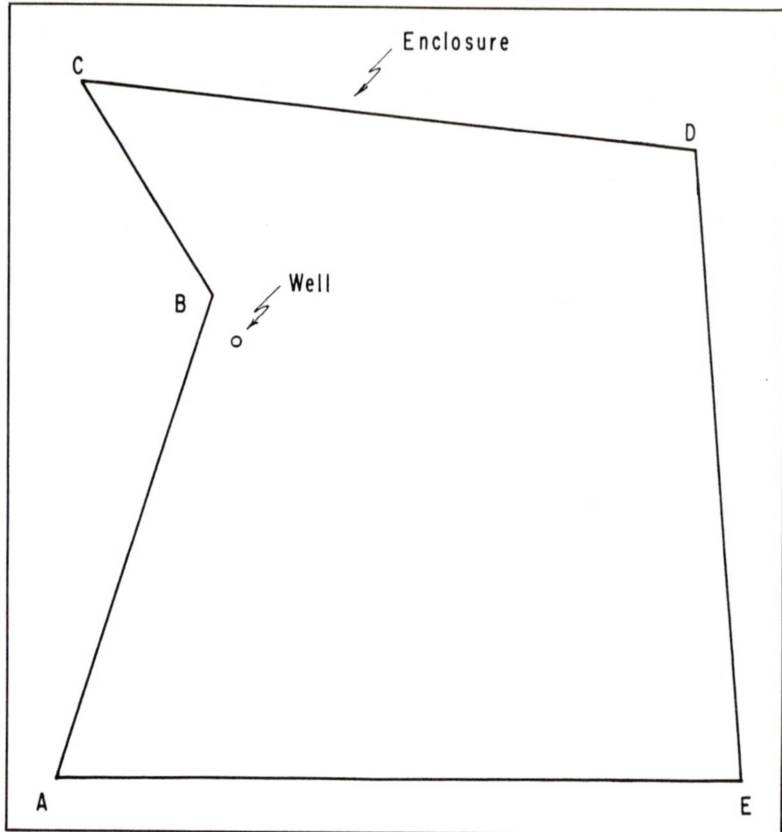

Fig. 1-1 Illustration of a structural fault block.

MODELS

It has been discussed in an earlier paper[1], and presented in Chapter 2 of this book, that what is reported in the literature for conformal mapping of a circle is inapplicable to our needs, because it involves a singularity of the origin representative for radial flow.

We have to map the entire domain within the circle free of any such singularities if we are to learn the feasibility of this method.

The reason and choice for considering a circle and then transforming this to the plane of a rectangle is that we know all the solutions that apply to a well within a circle, whether for steady-state or transient fluid flow, and this offers a check on the method of conformal mapping.

Fig. 1-2 and 1-3 are the graphical illustrations of these models, especially derived for this text. (See Mathematical Derivations at the end of this chapter.) These solutions were undertaken because graphical representations could not be found in the literature.

Fig. 1-2 represents a point source and sink opposite to one another on the circumference of a circle, with the potentials and streamlines so developed mapped in the interior.

It was learned very early in this investigation that such a model could not be used, because here again we were witnessing singularities at the source and sink that made it inappropriate for any conformal presentation.

From this it was learned how to apply the line source and sink placed on the circumference, opposite to one another, each subtending 45°.

This model, Fig. 1-3, representing a radius of unity for the circle, with its potentials and streamlines mapped within, was the basis for reproducing this configuration to a rectangle to perform the mathematics that applied to Equation 1-1.

CONFORMAL REPRESENTATION

The analogy for transforming this configuration to a rectangle is electricity and the application of Ohm's law.

The choice of the circle that we wished to investigate, represented an exterior radius of $r_e = 100$, with the well, wherever placed in this domain, having a radius of unity.

Thus, the area involved is

$$A = \overline{100}^2 \times \pi = 31{,}415.92654$$

The resistance of this model read from Fig. 1-3 is given by,

$$\frac{\nabla\phi}{\nabla\psi} = \frac{1.0000}{0.708871}$$

$$= 1.410694$$

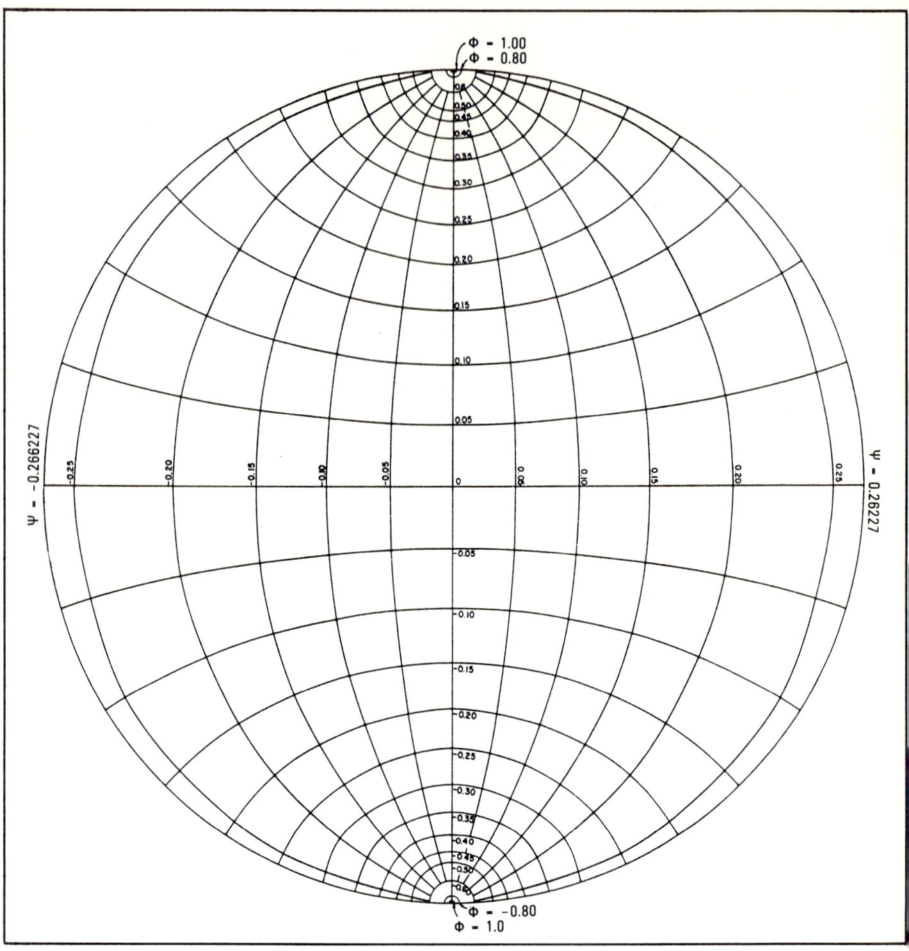

Fig. 1-2 Potential and streamline distributions within a circle for the point source and sink located on the circumference.

The resistance of the rectangle that follows from the resistance of a cable,—proportional to its length divided by the cross-sectional area —but now applied to a plane, is

$$\frac{L}{W} = 1.410694$$

The resistivity is unity, if one follows through the mathematics.

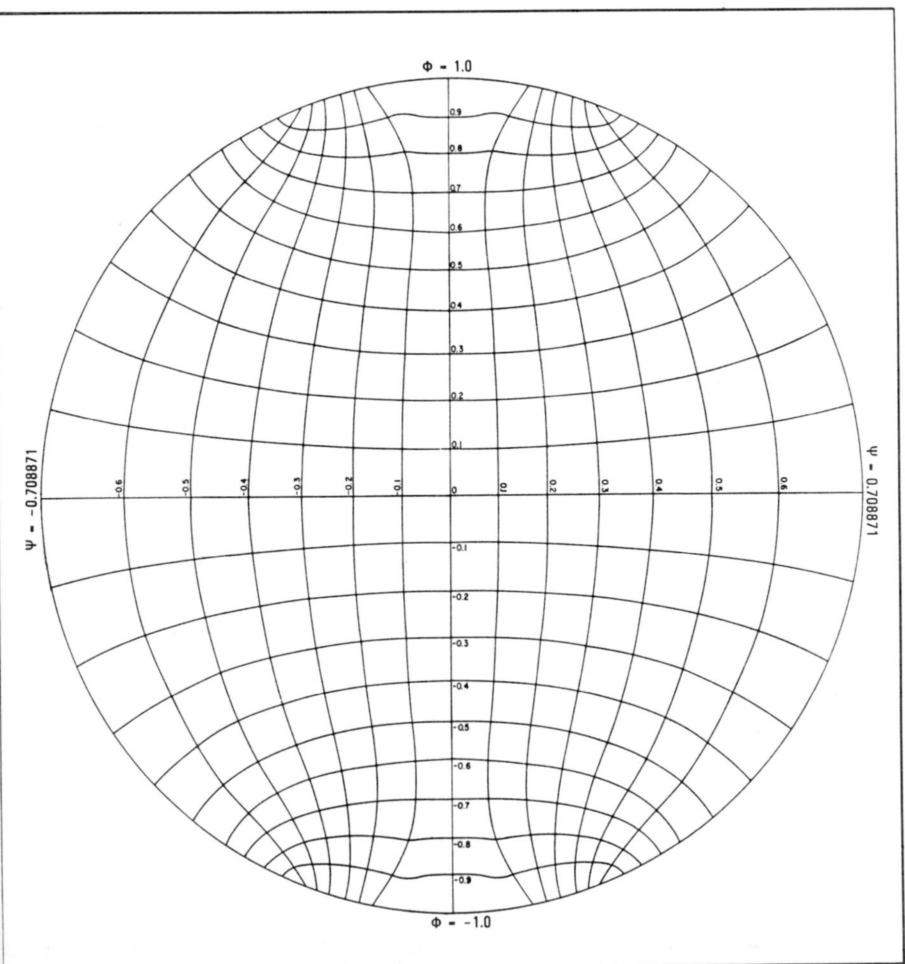

Fig. 1-3 Illustration of potential and streamline distributions within a circle. The line source and sink are located on the circumference opposite one another, and each subtends 45°.

Therefore

$$A = 31{,}415.92654 = L \times W$$
$$= 1.410694W^2$$

and,

$$W = 149.230821$$

with

$$L = 210.519022$$

This is the conformal rectangle shown in Fig. 1-4, with all the potentials and streamlines reproduced linearly from Fig. 1-3, with $\phi = 1.00$ for $(\bar{x}, \bar{y}) = (0, 105.25951)$, and $\Psi = 0.708871$ for $(\bar{x}, \bar{y}) = (74.61541, 0)$, conforming to the pattern of the circle that is the model.

TRANSIENT FLUID FLOW

To develop the engineering application and its significance to the reader, we wish first to determine the pressure change with time for a point located at $(x, y = (100, 0)$ on the circle with the well placed at the origin. This corresponds to the point on the rectangle of $(\bar{x}, \bar{y}) = (74.61541, 0)$ in Fig. 1-4, and the origin of the rectangle is the location of the well.

The interpretation of transient fluid flow for a point on the rectangle is the application of the *Ei* functions for which tables are available in the literature[2]. Thus to express the pressure drop in an infinite medium for the flow of fluids

$$\Delta p = \frac{q\mu}{2\pi kh} \times \frac{1}{2}\left\{-Ei\left(\frac{-r^2}{4t_D}\right)\right\} \tag{1-2}$$

and what we shall consider in this formula, is the $P(r, t_D)$ function, which is

$$P(r,t_D) = \frac{1}{2}\left\{-Ei\left(\frac{-r^2}{4t_D}\right)\right\} \tag{1-3}$$

With the well located at the origin in Fig. 1-4, this is taken as a series of image wells in the $\bar{x}, \bar{y}$ plane, duplicating the location of well

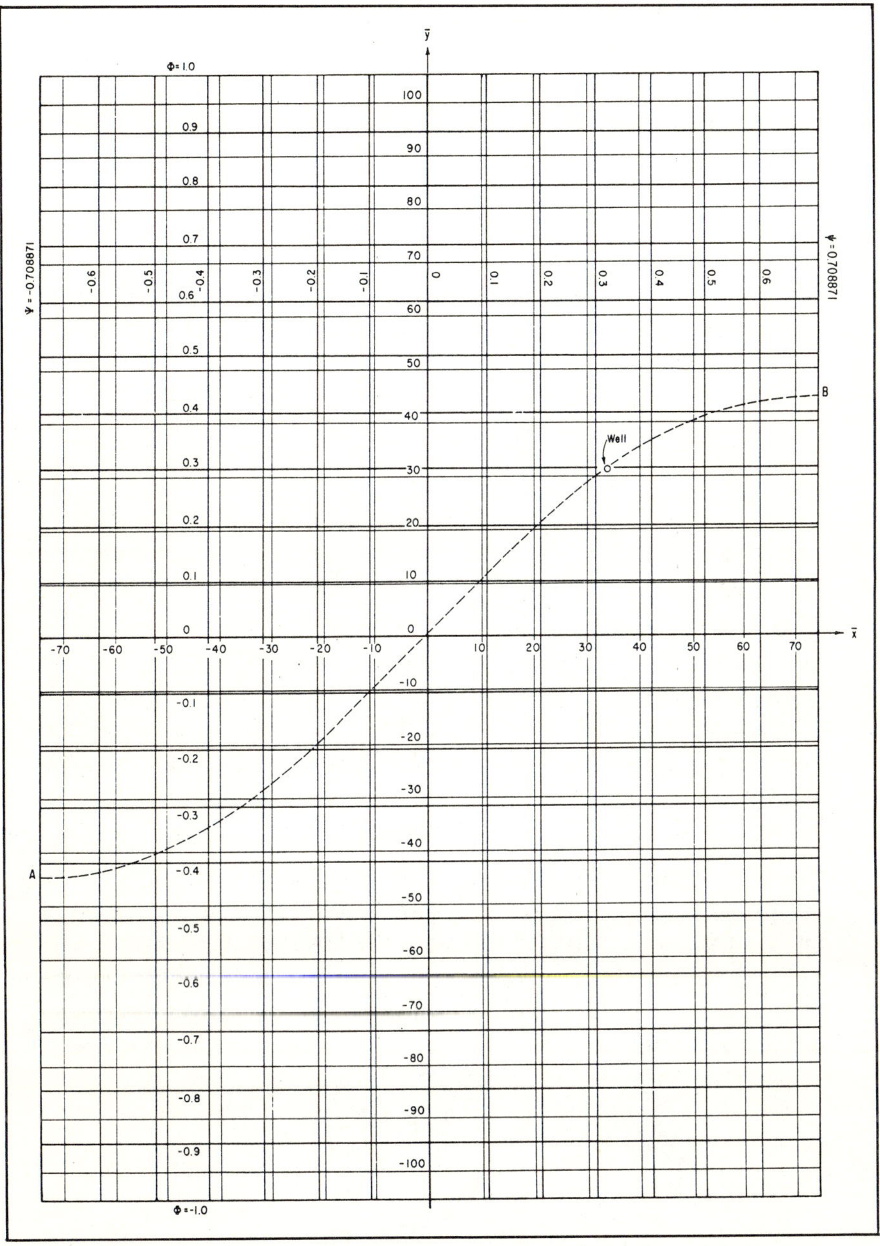

Fig. 1-4 Conformal rectangle reproduced for the circle, $r_e = 100$.

at the origin. This is done as a series of sinks, such that no fluid flows across the boundaries of the rectangle, but referring to the point in question $(\bar{x}, \bar{y}) = (74.61541, 0)$ as each of these sinks contribute to its pressure drop, $P(r, t_D)$.

The results are shown in Fig. 1-5. The upper line is what is established by conformal mapping. The second straight line is calculated from the work of Muskat[3] and Hurst[4] for a well located at the origin subject to a constant rate for an enclosed reservoir of r_e = 100, with the point of reference at the boundary of the circle.

Also involved is a scaling problem due to the radius of drainage and the time it takes to reach the point of drainage[5, 6]. The drainage distance to the boundary of the rectangle is 74.61541, and for the circle it is $re = 100$, the shortest distance from a well to the point of reference.

Thus, by applying the drainage formula

$$r_d = 2.6408\sqrt{t_D} \tag{1-4}$$

to determine t_D for each of these distances, the difference in times is the shift on the t_D axis for the $P(t_D)$ function established by conformal mapping.

The addition or subtraction of a time constant does not inviolate the precepts of Equation 1-1.

Fig. 1-6 is for the same well located at the origin, but here the function applies for the well radius at unity. The match in this respect is almost perfect. There is no shift in the time scale, because what is unity for the well radius on the circle is the same on the rectangle.

The author realized it was necessary to investigate points within the interior of Fig. 1-3 to observe if these results are sustained.

Such is represented as an off-center well for a line drawn 45° through the origin of Fig. 1-3. This would place the well at $r = 50$, and $\theta = 45°$, and on the rectangle Fig. 1-4 this would correspond to the point $(\bar{x}, \bar{y}) = (33.864300, 30.391999)$. The areal projection is reproduced in Fig. 1-4 as the curve AB.

The reference is the well-bore radius of unity for this off-center well. Its comparison and check with Muskat's solution[3] for his Green Function Theorem shows excellent agreement in Fig. 1-7.

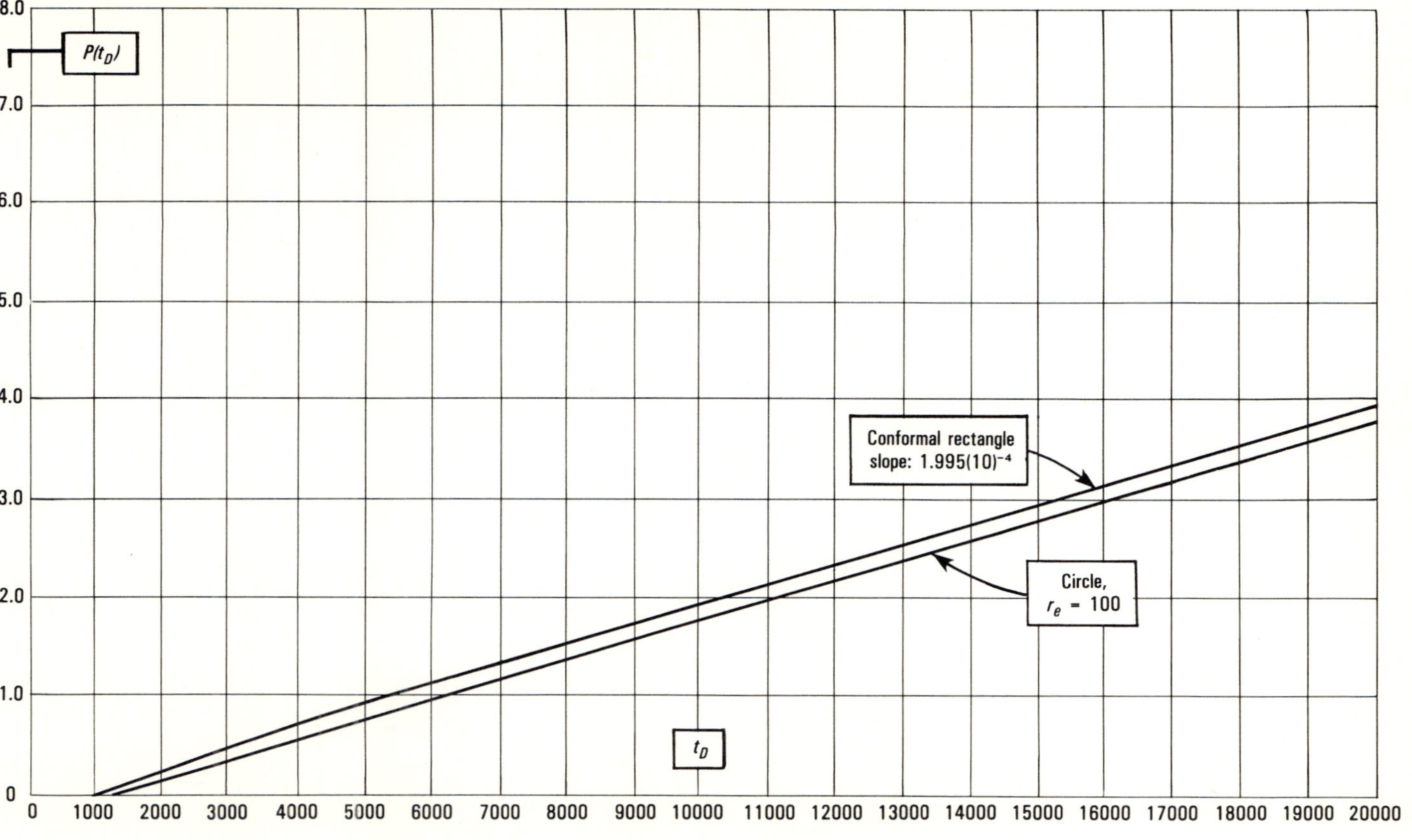

Fig. 1-5 *Well at origin, $P(t_D)$ calculated for $r_e = 100$, and $\theta = 0$.*

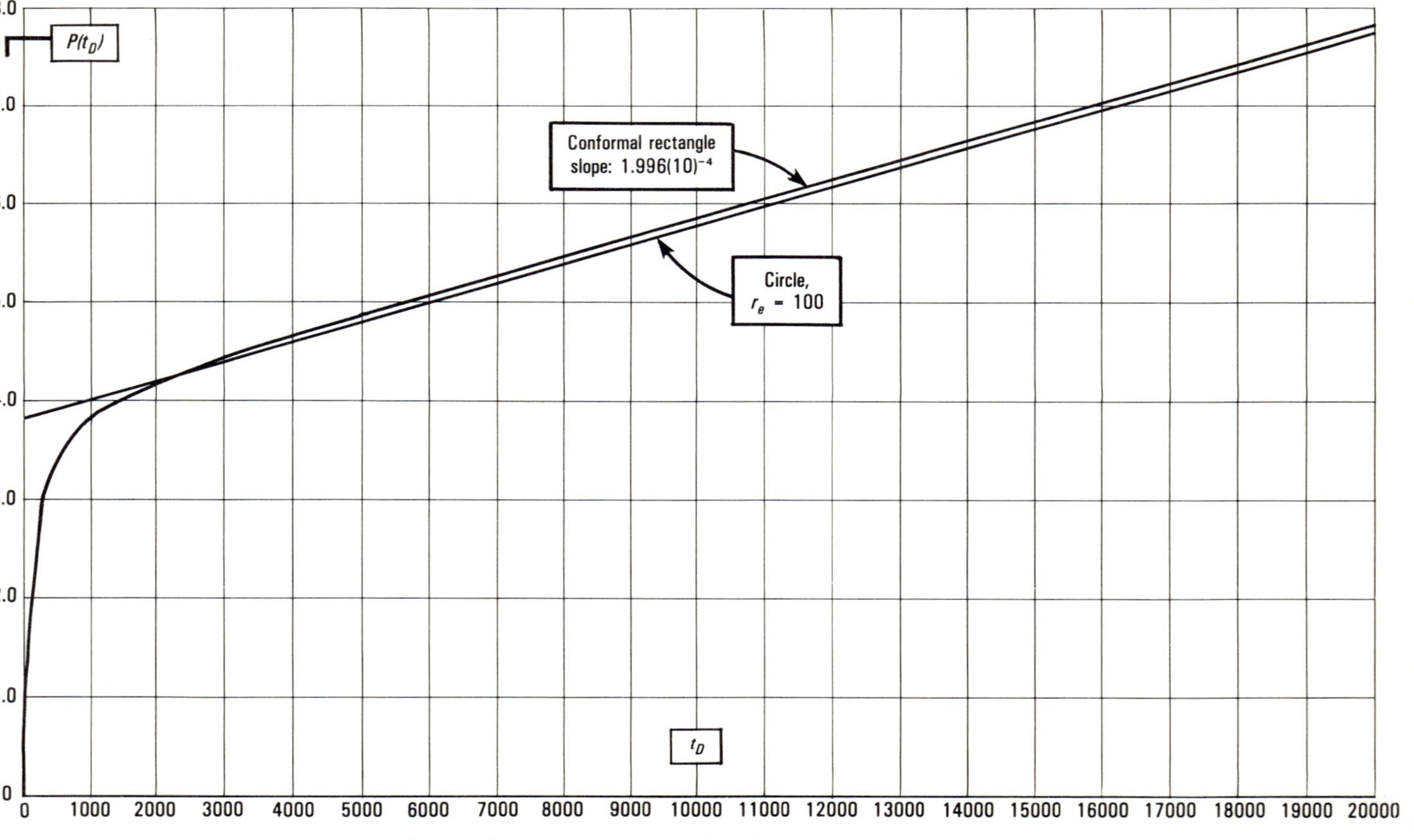

Fig. 1-6 $P(t_D)$ *determined for well radius unity, well at the origin.*

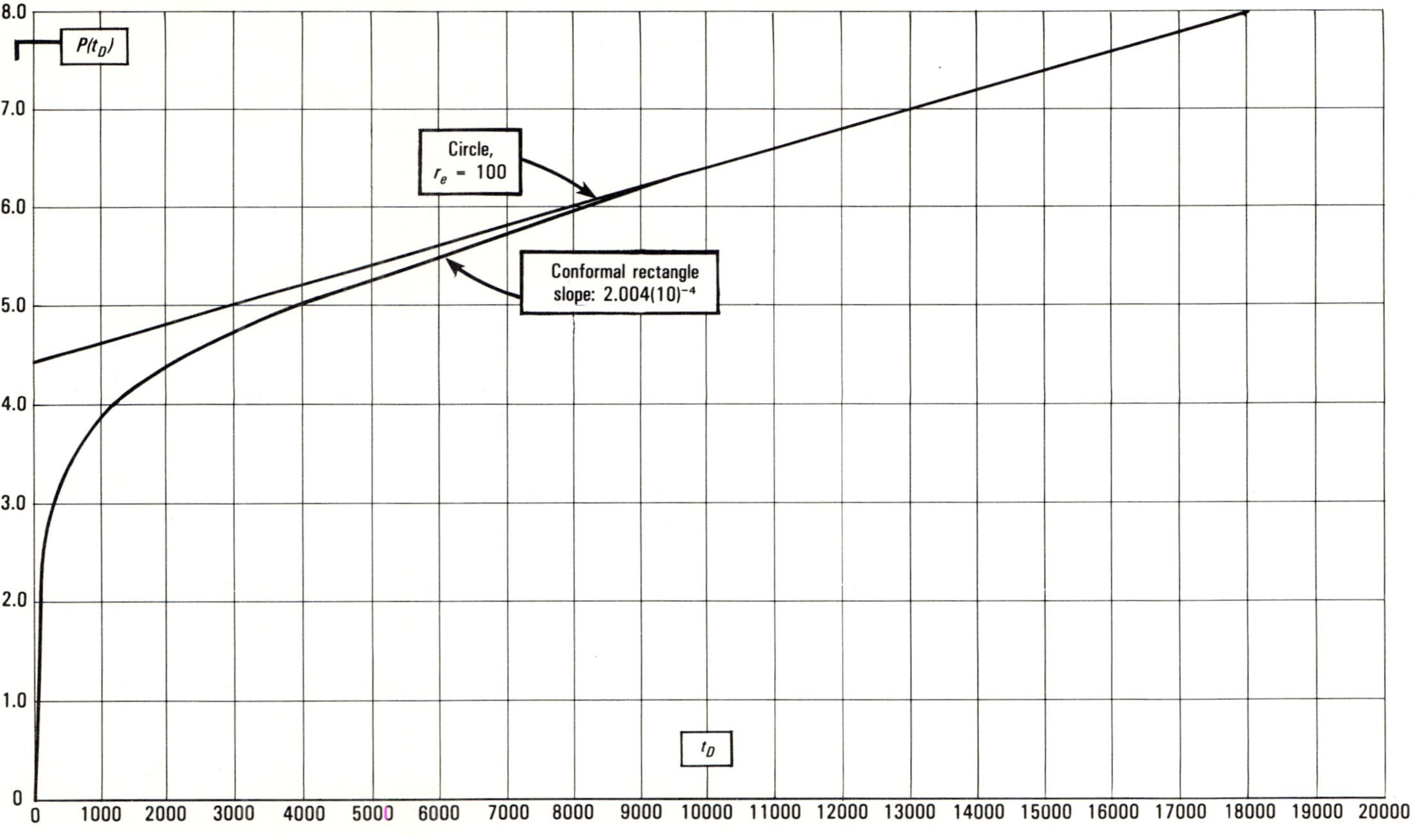

Fig. 1-7 Off-center well at $r = 50$ and $\theta = 45°$, $P(t_D)$ established for well radius unity.

For the other extreme corresponding to this line for the point r_e = 100 and θ = 225°, Fig. 1-8, shows the close comparison for the $P(t_D)$ function realized with the Muskat solution and conformal mapping.

In all these illustrations the shift in the time scale has been positive, because the distances of the radii on the reference rectangle have always been less than the corresponding distances that applied to the circle.

We wished to observe a negative shift, or the opposite effect. This is shown in Fig. 1-9, where the well is again located at the origin, and the point of reference is r_e = 100, and θ = 67.5°. This distance on the rectangle is 129.023346 that was reduced to r_e = 100 by the method of scaling.

Finally, the reader would ask why these applications are limited to a small radius of r_e = 100 for a circle, and not to a large and extended radius that would apply for the circle. This is shown in Fig. 1-10 for r_e = 3,000.

The summary of the cases illustrated for a center well represented on the OX-axis in Fig. 1-3, and the off-center well for a line drawn 45° through the origin are shown in Figs. 1-11 and 1-12, respectively, and correspond to the time for t_D = 15,000.

These close agreements prove the adaptability of conformal mapping to reservoir performance.

Further, the slopes shown conform to $2\pi/A$ that attest to the application of using image wells, and show that the mathematics for a rectangle apply to a reservoir.

The readings of the Ei functions from tables, and their interpolations herein employed are not tedious but lengthy. For this reason, it is recommended that this process be programmed to facilitate the work.

There are new programmable calculators that seem most appropriate for this work.

In dealing with the *Ei* function this can be incorporated in the program as a subroutine. Its formula is given as

$$-Ei(-z) = -\gamma - In\, z - \sum_{n=1}^{Inf.} \frac{(-1)^n z^n}{n\, n!} \tag{1-5}$$

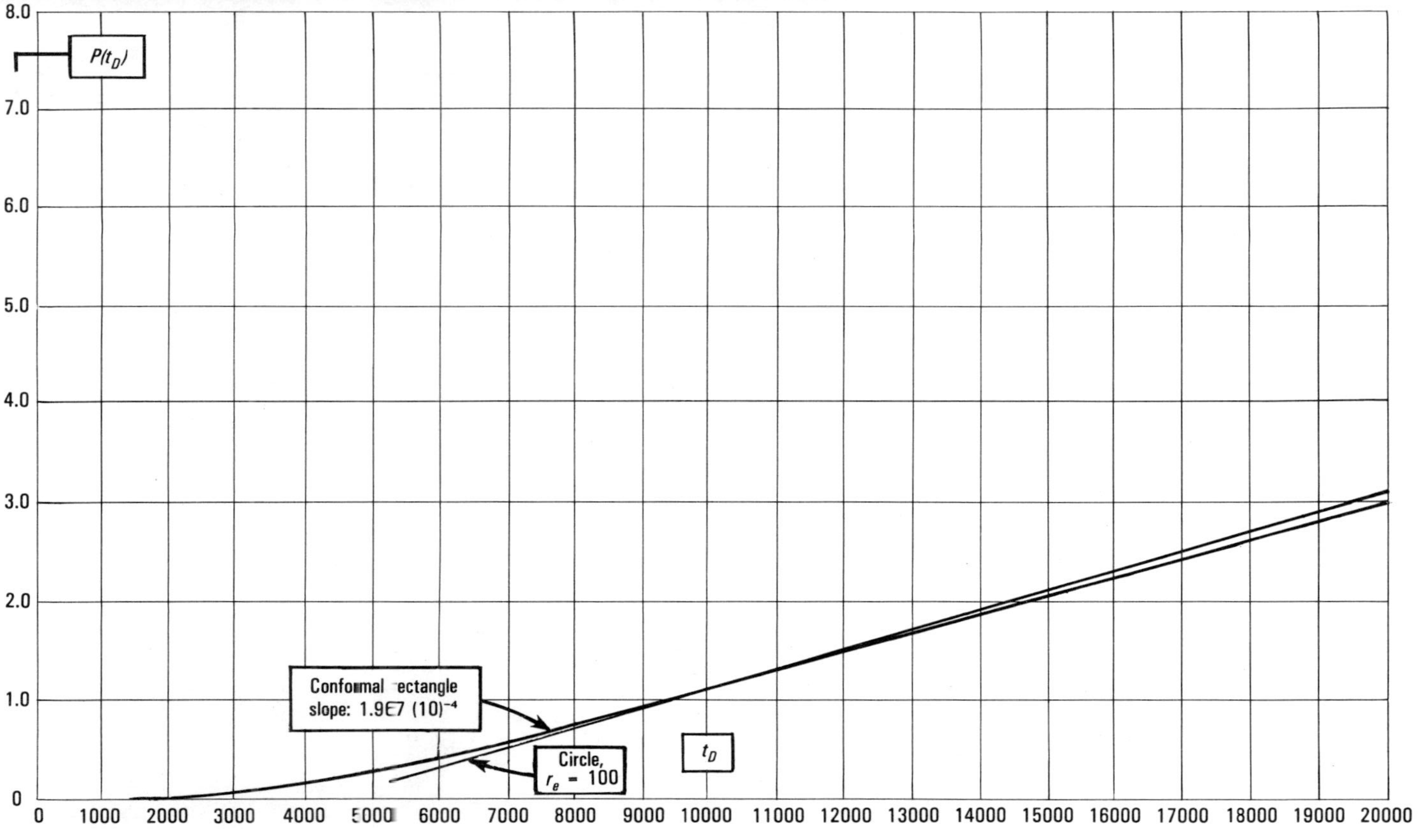

Fig. 1-3 Off-center well, $P(t_D)$ determined for r_e *= 100, and* θ *= 225°.*

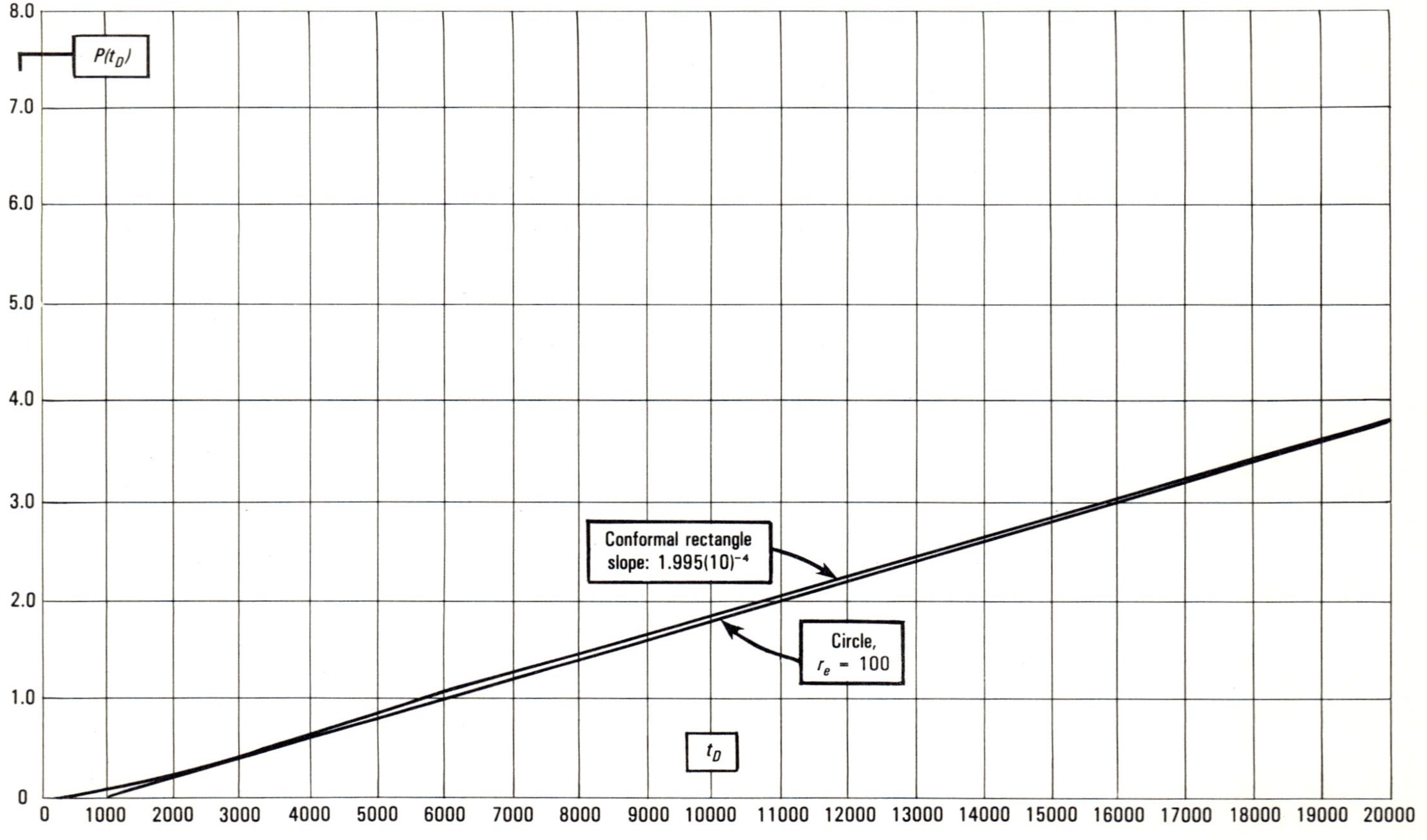

Fig. 1-9 Well at origin, $P(t_D)$ calculated for $r_e = 100$ and $\theta = 67.5°$.

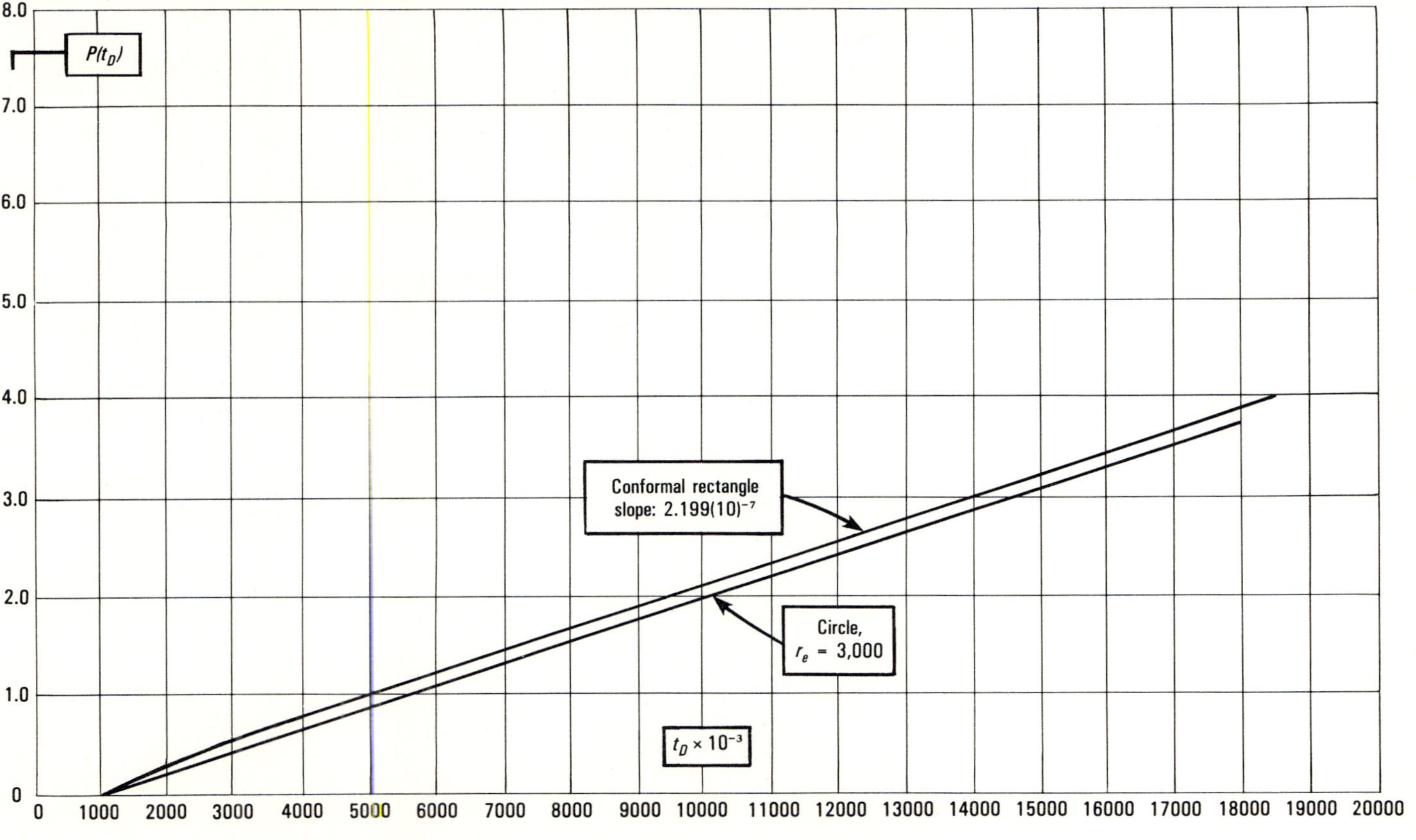

Fig. 1-10 Extended circle, $P(t_D)$ determined for $r_e = 3,000$, and $\theta = 0$, well at origin.

where γ is Euler's Constant, equal to 0.577216, the derivation is given in Bromwich[7]. See Table A-1 (Appendix) for calculation of $Ei(-Z)$ function.

For large values of z, the asymptotic expansion applies, referring to Knopp[8], where

$$-Ei(-z) = e^{-z}\left\{\frac{1}{z} - \frac{1}{z^2} + \frac{2!}{z^3} - \frac{3!}{z^4} \pm\right\} \qquad (1\text{-}6)$$

INSTRUMENTATION

The problem that presents itself is not a circle, but any configuration of a structural fault block or a field that the geologist chooses to place before us. We must now determine how to handle it using conformal mapping.

Fig. 1-1 offers a good example. Using an electricity analog, consider Fig. 1-1 an electrolytic trough containing an electrolyte such as copper sulfate solution. A constant voltage by means of a copper strip is placed across the side ABC, as the source, with another strip of lower but fixed voltage, across the side DE at the sink. By means of a probing conductor, all potentials and streamlines within this trough can be mapped yielding a plot comparable to what is shown in Fig. 1-3.

The conversion to a conformal rectangle follows exactly as given in this text.

This is known as electrolytic modeling, and the reference to a solid conductor is the potentiometric model.

At the present stage, this is almost a lost art and science to this industry, although Muskat[3] and the writer[9] have reported on this method in the past.

The second method is programming using the concept of Equation 1-1 for steady-state, with the boundaries defined by the line source and sink with the enclosures of the field.

The reiteration and relaxation procedures used for this purpose would require a network of meshes that could be performed with a digital or analog computer.

The final method that the writer can refer to is the Schwartz-Christoffel method. This is a mathematical approach, that offers

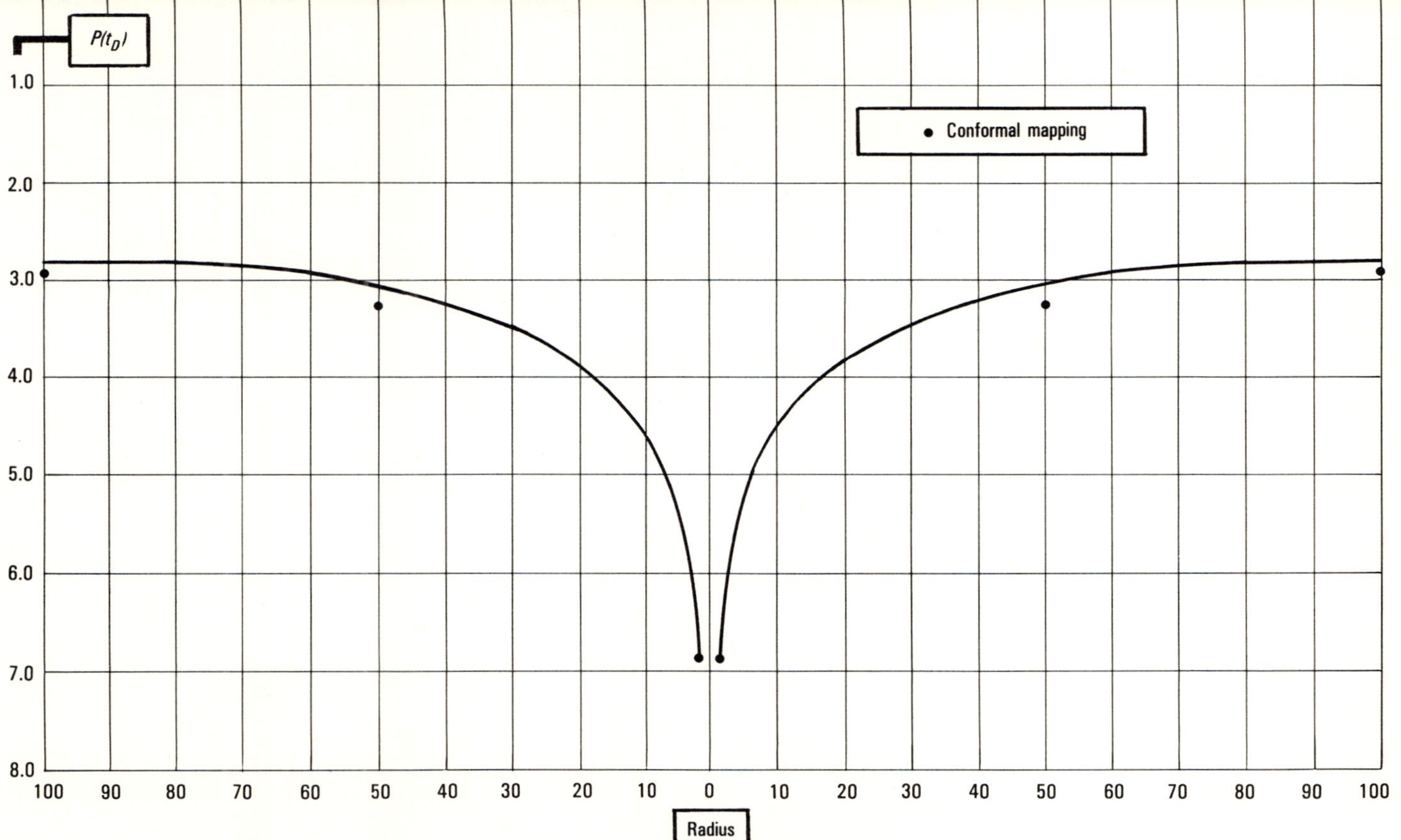

Fig. 1-11 *Plane through the OX-axis, $P(t_D)$ vs r for well at origin.*

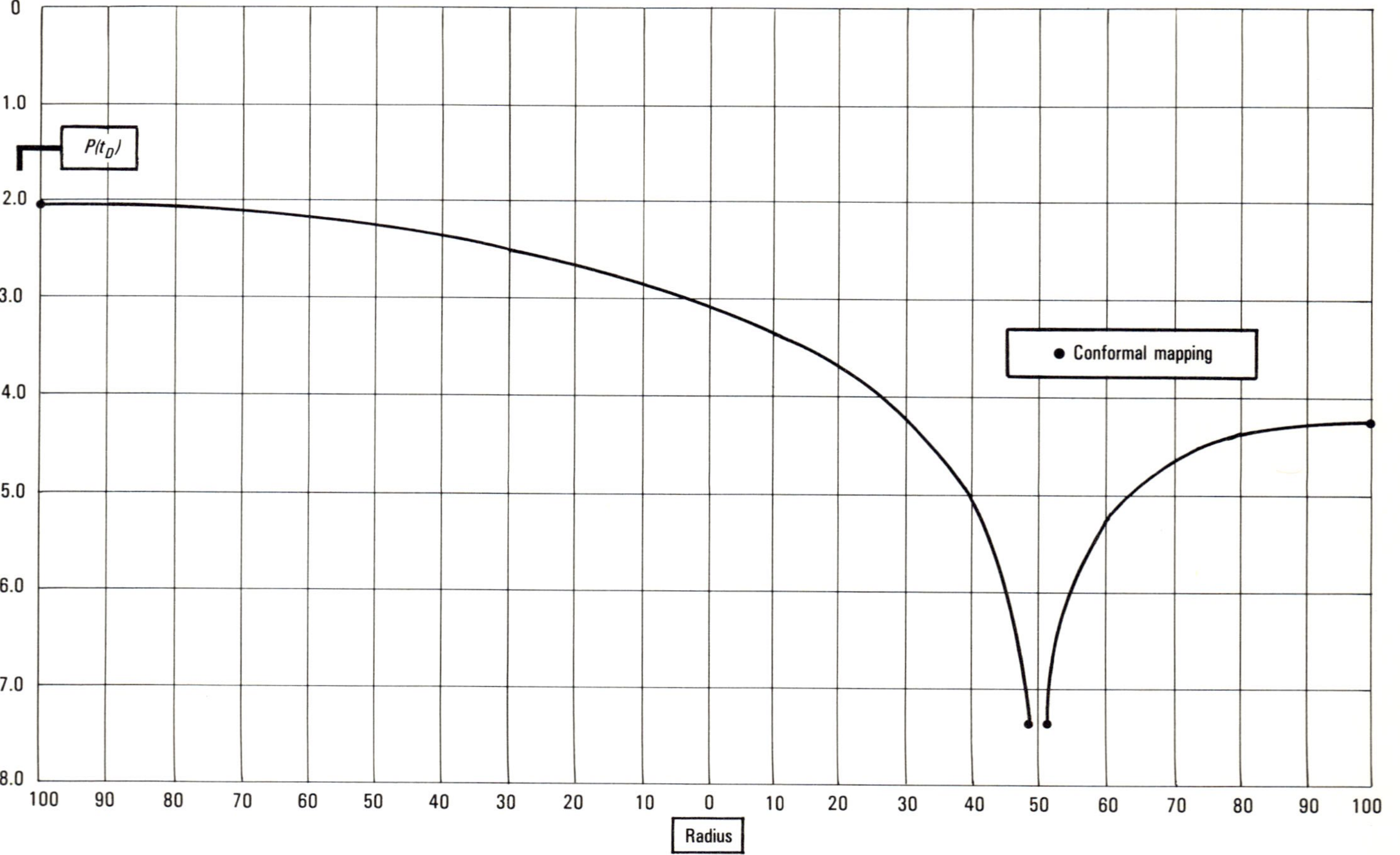

Fig. 1-12 $P(t_D)$ *vs r for off-center well, plane passed through the origin at* $45°$.

certain complexities in treating with the elliptical functions in its solution. The problems so treated as observed from the literature are limited, but this is in the province of the mathematician.

APPLICATION

The performance of a well itself will no longer be the idealized and uniform conditions of radial flow for an extended medium, but will take on the complexities of the bounded conditions illustrated in Fig. 1-1. The $P(t_D)$ values so derived would apply in any such bottom-hole pressure draw-down or build-up test. This is one reason for including transient fluid flow in this investigation.

These $P(t_D)$ values would also be applicable to consider the nonlinear solutions for the flow of gases, and the two-phase fluid flow of oil and its associated gas both flowing to a well and mapped within the field. These have been reported by the author[1, 10] and are repeated in Chapters 2 and 3 of this book.

These plots as illustrated are also general-purpose plots.They not only apply for transient fluid flow, but for the semisteady-state flow when all the boundaries are effected and pressure depletion sets in the field.

This applies to any location of wells or a well within a reservoir. The conformal mapping once performed can be used for transient fluid flow or steady-state flow for the well located in an enclosed boundary represented by the rectangle.

The depicting of the steady-state flow within an enclosure was reported by Basset[11] at the turn of the century. Muskat[3] and others have employed this to represent waterfloods and gas-cycling patterns that are reported in the literature.

Thus, semisteady state during the depletion can be developed on the rectangle to yield its pressure and streamline distributions in the reservoir.

Its application by these streamlines would be to define where bodies of oil have been entrapped, and the need to recover this oil by in-fill drilling. Such fields known to exist in Louisiana have produced for the last 30 years.

The application of these streamlines from injection to producing wells can be applied to tertiary recovery where surfactant treatments are used in the Mid-Continent.

Also, in a distorted pattern as shown in Fig. 1-1, the drainage across a lease line or within a field can only be performed by conformal mapping[12].

These are some of the innovations that follow from conformal mapping.

NONCONTINUOUS FORMATIONS

The problem is not simply the structural fault block illustrated in Fig. 1-1, but within this domain the geologist locates faults and shale lines that are barriers to fluid flow.

An electrolytic model such as discussed for Fig. 1-1 can be used to represent this condition by inserting insulated barriers for the faults and shale lines within the electrolyte. What would result in reproducing this in the conformal rectangle, however, are unmapped areas that would be intractable for solution.

The idea is to obtain a plat such as Fig. 1-4 that is continuous in the plotting of potentials and streamlines in the rectangle.

What has been conceived by the writer is to make these barriers within the reservoir as sources and sinks, identified with the potentials for the enclosures. This is illustrated in Fig. 1-13 which is adapted from Fig. 1-1.

Thus by using copper strips for the faults within the electrolyte, charged with a voltage of unity, and another copper strip for the shale line with a voltage of minus unity, these discontinuities then become identified with the boundaries for the rectangle, Fig. 1-4, for the potentials of plus and minus unity.

Its conformal rectangle would then be applicable to water or tertiary flooding to show the paths of the streamlines around these barriers as the injected fluid progresses to reach the producing wells.

The $P(t_D)$ function so established for any well location in the field applies for well test performance, as it has been shown in Figs. 1-6 and 1-7 that these $P(t_D)$ values are absolute—there is no time shift involved.

For points in the open domain relative to the well in this complex

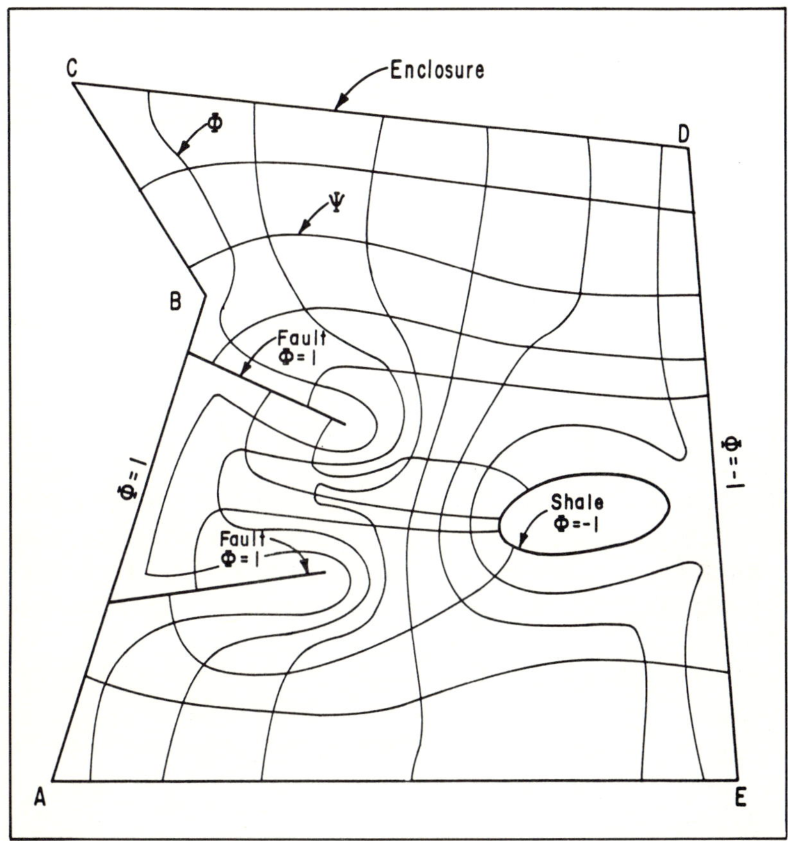

Fig. 1-13 Illustration of a structural fault block including fault barriers and shalelines.

structure the time shift hereto described is in effect; however, for points removed from the well behind a fault or barrier, yet influenced by the voidage from the well, the radius-of-drainage formula, Equation 1-4, can no longer apply.

Another approach is necessary for the solution of this problem that is equivalent to the Green Function often referred to in the literature.

This is the application of the Basset formula, and the determination of semisteady-state flow during the depletion of the field, where the potential at the well is taken as zero, for which all other potentials within the rectangle are established relative to this value.

The reference is the well itself and the straight-line portion of its $P(t_D)$ values, for which a time point is chosen.

Thus, to apply this method to any point within the structure, its value established by semisteady-state is subtracted from the ordinate value of the reference point, and a straight line with a slope of $2\pi/A$ is drawn through the point.

This is the asymptotic value that the $P(t_D)$ function for a distant point will approach and would represent its time shift.

This has been applied to the problem for the off-center well in Fig. 1-4, performing the semisteady-state calculations within the rectangle to show the close agreement with the illustrated curves in Fig. 1-12. See Table A-2 in Appendix for Muskat's solution.

MATHMEMATICAL DERIVATIONS

For a point located within the domain of a circle, Fig. 1-2, its pressure drop with respect to a source placed on the circumference, is

$$\phi = -In_1r_2$$

where ${}_1r_2$, is the distance from the source to the point.

For a sink located on the same circumference, its pressure change is the relation

$$\phi = In_2r_3$$

where the distance ${}_2r_3$, is the measurement of the distance from the point to the sink.

Since the pressure at the point is influenced by both the source and the sink, these effects are additive, and

$$\phi_p = In_2r_3 / {}_1r_2$$

or expressed in polar coordinates, gives

$$\phi_p = q\ In\ \frac{(1 + 2\bar{r}\ cos(\theta^1 - \theta) + \bar{r}^2)}{(1 - 2\bar{r}\ cos(\theta^1 - \theta) + \bar{r}^2)} \tag{1-7}$$

with q, the coefficient of fluid flow, and
$\bar{r}$ equals r/r_e, and all radii are measured relative to a circle of unit radius.

The location of the point, therefore, is $(\bar{r}, \theta)$ within the circle and the reference to the source is (1, 90°), where $\theta^1 = 90°$.

The relationship as shown in Eq. 1-7, is contained within an enclosed circle. The reader can verify this by differentiating Eq. 1-7 with respect to $\bar{r}$, and let this approach unity to show that this slope is zero at the boundary.

However, since Equation 1-7 can go to infinity at the source or at the sink, there has to be a "cut-off" point. Therefore what has been chosen for $\phi = 1$ at $(\bar{r}, \theta) = (0.995, 90°)$, yields $q = 0.0834868$.

In treating with the streamline development, cartesian coordinates are more applicable for this determination.

Therefore, Equation 1-7 can also be expressed as

$$\phi_p = q \; In \; \frac{x^2 + (y + 1)^2}{x^2 + (y - 1)^2} \tag{1-8}$$

Further, by the classical interpretation of a streamline

$$\frac{\partial \psi}{\partial x} = -\frac{\partial \phi}{\partial y} \tag{1-9}$$

it follows that

$$\frac{\partial \phi_p}{\partial y} = -q \left\{ \frac{2(y + 1)}{x^2 + (y + 1)^2} - \frac{2(y - 1)}{x^2 + (y - 1)^2} \right\} \tag{1-10}$$

Therefore, by Equation 1-9

$$\begin{aligned} \psi &= -2q \int_0^x \left\{ \frac{(y + 1)}{x^2 + (y + 1)^2} - \frac{(y - 1)}{x^2 + (y - 1)^2} \right\} dx \\ &= -2q \left[\tan^{-1} \frac{2x}{1 - x^2 - y^2} \right] \end{aligned} \tag{1-11}$$

Since we wish to make the streamlines positive that would apply in the first quadrant to establish the conformal rectangle, the sign in Equation 1-11 is omitted, which is shown in Fig. 1-2.

The line source solution illustrated in Fig. 1-3 follows from the foregoing. This is the integration of the point-source solution as a continuous function along the circumference of the circle, expressed as

$$[\phi(\bar{r},(\bar{\theta}_2-\theta)) - \phi(\bar{r},(\bar{\theta}_1-\theta))]$$

$$= \int_{\bar{\theta}_1}^{\bar{\theta}_2} ln \frac{(1+2\bar{r}\cos(\theta^1-\theta)+\bar{r}^2)}{(1-2\bar{r}\cos(\theta^1-\theta)+\bar{r}^2)} \, d\theta^1 \tag{1-12}$$

$$= \int_{\bar{\theta}_1-\theta}^{\bar{\theta}_2-\theta} ln \frac{(1+2\bar{r}\cos u+\bar{r}^2)}{(1-2\bar{r}\cos u+\bar{r}^2)} \, du$$

As shown in that illustration the subtending angle for the source and the sink is 45° on the circumference. This was arbitrarily set to get the best coverage for the mapping of potentials and streamlines within the circle. The results indicate there is no limitation placed on the length and size of a potential placed in an electrolytic model to map an interior. The criterion is to obtain the best coverage for mapping.

In generating a point source along a line or arc, experience indicates a constant potential cannot be determined over any large distance. If we use segments as in Equation 1-12 with their respective coefficients of fluid flux determined, however, the constant potential can be established.

This is the procedure used in the development of Fig. 1-3, dividing the source as well as the sink in three separate arcs of 15° each, and oriented symmetrically to the 90° and 270° positions on the circle for the source and the sink.

To accomplish this, the integration of Equation 1-12 had to be developed. Such could be established analytically as a series expansion in $\bar{r}$ and $sin(\bar{\theta}-\theta)$; however, it was learned early that this was a poorly convergent series, and divergent for $\bar{r} = 1$. The other alternative and practical approach was to carry out this integration numerically, Table 1-1.

The table covers increments of 15° for $\bar{\theta} - \theta$, and for $\bar{r}$ varying from zero to unity.

Intermediate values in angular displacement were obtained by the Four-Point Lagrangian Interpolations,[13] and for negative values in $\bar{\theta} - \theta$, the recurrence formula applied

$$\phi(\bar{r}, -(\bar{\theta}-\theta)) = -\phi(\bar{r}, (\bar{\theta}-\theta)) \tag{1-13}$$

This concept of separate segments to represent a line source and employing the numerical values for $\phi(1, (\bar{\theta}-\theta))$, the coefficient for fluid flux could be calculated algebraically for selective points along the source.

This showed fairly consistent values for $\phi = 1$ at the source, but at the terminal points ϕ evidenced a 6% lowering from unity.

The expediency as the writer recognized was to add the point source solution shown earlier to each of the extremities of the line

TABLE 1-1
Numerical integration of Equation 1-12, ϕ ($\bar{r}$, $(\bar{\theta}_2-\theta)$) vs $(\bar{\theta}-\theta)$

$\bar{\theta}-\theta$	$\bar{r}=0.2$	$\bar{r}=0.4$	$\bar{r}=0.6$	$\bar{r}=0.8$	$\bar{r}=0.85$	$\bar{r}=0.90$	$\bar{r}=0.95$	$\bar{r}=1.00$
0	0	0	0	0	0	0	0	0
15°	0.209621	0.435944	0.703564	1.060456	1.173079	1.298053	1.436150	1.587838
30°	0.403584	0.829196	1.300456	1.838448	1.984075	2.133528	2.286630	2.443167
45°	0.568164	1.150256	1.754848	2.380080	2.538092	2.696071	2.854042	3.012114
60°	0.692780	1.384352	2.069416	2.739235	2.902985	3.064829	3.224976	3.383729
75°	0.770284	1.525836	2.253718	2.945226	3.111662	3.275299	3.436451	3.595517
90°	0.796540	1.573092	2.314389	3.012444	3.179675	3.343844	3.505293	3.664452
105°	0.770284	1.525836	2.253718	2.945226	3.111662	3.275299	3.436451	3.595517
120°	0.692780	1.384352	2.069416	2.739235	2.902985	3.064829	3.224976	3.383729
135°	0.568164	1.150256	1.754848	2.380080	2.538092	2.696071	2.854042	3.012114
150°	0.403584	0.829196	1.300456	1.838448	1.984075	2.133528	2.286630	2.443167
165°	0.209621	0.435944	0.703564	1.060456	1.173079	1.298053	1.436150	1.587838
180°	0	0	0	0	0	0	0	0

source and sink to yield the values for ϕ equal to 1.000, 1.000, 0.994696, 1.000, 0.994696, 1.000, 1.000 and for θ equal to 112.5, 105, 97.5, 90, 82.5, 75, and 67.5ϑ, respectively, a consistent check in the interpretation for ϕ.

Thus the formula employed for the potential distribution in Fig. 1-3, is the equation.

$$\begin{aligned}\phi = {} & 0.00914901\ \phi_p(\bar{r},(112.5^0-\theta)) \\ & + 0.442039\ [\phi(\bar{r},(112.5^0-\theta))-\phi(\bar{r},(97.5^0-\theta))] \\ & - 0.0920873\ [\phi(\bar{r},(97.5^0-\theta))-\phi(\bar{r},(82.5^0-\theta))] \\ & + 0.442039\ [\phi(\bar{r},(82.5^0-\theta))-\phi(\bar{r},(67.5^0-\theta))] \\ & + 0.00914901\ \phi_p(\bar{r},(67.5^0-\theta))\end{aligned} \qquad (1\text{-}14)$$

The streamline relationship expressed in polar coordinates is given as

$$\psi = \int_o^{\bar{r}} \frac{1}{\bar{r}} \frac{\partial\phi}{\partial\theta}\, d\bar{r} \qquad (1\text{-}15)$$

The integrand of Equation 1-15 offers the essential solution. As far as carrying out the integration with respect to $\bar{r}$, Simpson's rule applies.

Thus by Equations 1-7 and 1-12, introducing the coefficients of Equation 1-14, then

$$\frac{1}{\bar{r}}\frac{\partial\phi}{\partial\theta} = \frac{0.0365960\ \sin(112.5^0-\theta)(1+\bar{r}^2)}{(1+2\bar{r}\cos(112.5^0-\theta)+\bar{r}^2)(1-2\bar{r}(112.5^0-\theta)+\bar{r}^2)} -$$

$$\frac{+0.442039}{\bar{r}}\ In\frac{(1-2\bar{r}\cos(112.5^0-\theta)+\bar{r}^2)}{(1+2\bar{r}\cos(112.5^0-\theta)+\bar{r}^2)} \qquad (1\text{-}16)$$

$$x\ \frac{(1+2\bar{r}\cos(97.5^0-\theta)+\bar{r}^2)}{(1-2\bar{r}\cos(97.5^0-\theta)+\bar{r}^2)}$$

$$\frac{-0.0920873}{\bar{r}} \; ln \, \frac{(1-2\bar{r}\cos(97.5^0-\theta)+\bar{r}^2)}{(1+2\bar{r}\cos(97.5^0-\theta)+\bar{r}^2)}$$

$$x\,\frac{(1+2\bar{r}\cos(82.5^0-\theta)+\bar{r}^2)}{(1-2\bar{r}\cos(82.5^0-\theta)+\bar{r}^2)}$$

$$\frac{+0.442039}{\bar{r}} \; ln \, \frac{(1-2\bar{r}\cos(82.5^0-\theta)+\bar{r}^2)}{(1+2\bar{r}\cos(82.5^0-\theta)+\bar{r}^2)}$$

$$x\,\frac{(1+2\bar{r}\cos(67.5^0-\theta)+\bar{r}^2)}{(1-2\bar{r}\cos(67.5^0-\theta)+\bar{r}^2)}$$

$$\frac{+\,0.0365960\,\sin(67.5^0-\theta)(1+\bar{r}^2)}{(1+2\bar{r}\cos(67.5^0-\theta)+\bar{r}^2)(1-2\bar{r}\cos(67.5^0-\theta)+\bar{r}^2)}$$

This equation is involved as it is made up of several parts. However, its numerical application is fairly straightforward.

One problem did arise; namely, the convergence of the slope of the potential for $\bar{r} = 0$ at the origin. This was solved by the application of L'Hospital Rule for the determination of limits. Thus the slope of the potential with respect to $\bar{r}$ at the origin, established analytically gave this value as 0.863168. The differential of Φ with respect to $\bar{r}$ determined numerically along the OY-axis evidenced this to be 0.854734, a close comparison.

Finally, the sustainment of Equation 1-15 is summarized as follows: For radii drawn through $\theta = 0$, 15, 30, 45 and 60° at the origin, the calculated ψ's at $\bar{r} = 1$, are 0.708871, 0.708870, 0.708873, 0.708868, and 0.708492, respectively.

REFERENCES

1. William Hurst, "The Solution of Non-Linear Equations for Flow of Fluids," Parts I and II. THE OIL AND GAS JOURNAL, December 10, 1973 and December 17, 1973, respectively.

2. Tables of Sine, Cosine, and Exponential Integrals, Vols. I and II, Federal Works Agency Work Projects Administration, sponsored by National Bureau of Standards, 1940.

3. M. Muskat, FLOW OF HOMOGENEOUS FLUIDS, J. W. Edwards, Inc., 1946.

4. A. F. van Everdingen and W. Hurst, "The Application of the Laplace Transformation to Flow Problems in Reservoir," Trans. AIME, Vol. 186, 1949.

5. William Hurst, Orville K. Haynie and Richard W. Walker, "New Concepts Extends Pressure Build-Up Analysis," PETROLEUM ENGINEER, August, 1962: Based on paper, "Some Problems in Pressure Build-Up," SPE, 145, by the cited authors, presented at the 1961 Fall Meeting of the AIME, Dallas, Texas.

6. William Hurst, "The Radius of Drainge Formula," THE OIL AND GAS JOURNAL, July 14, 1969.

7. T. J. I'A Bromwich, AN INTRODUCTION TO THE THEORY OF INFINITE SERIES. Macmillian and Co., Limited, London, 1949, p. 334.

8. Konrd Knopp, THEORY AND APPLICATION OF INFINITE SERIES, Blackie & Son, Limited, London and Glasgow, 1928, p. 549.

9. William Hurst and G. M. McCarty, "The Application of Electrical Models to the Study of Recycling Operations in Gas-Distillate Fields," Drilling and Production Practice, API, 1942.

10. William Hurst, "Subsidiary Equations," Parts I and II, THE OIL AND GAS JOURNAL, October 8, 1973, and October 15, 1973, respectively.

11. A. B. Basset, HYDRODYNAMICS, Dover Publications, Inc., 1961.

12. William Hurst, "How to Figure Oil Drainage Across a Lease-Line," THE OIL AND GAS JOURNAL, January 20, 1975.

13. Tables of Lagrangian Interpolation Coefficients, National Bureau of Standards, Columbia University Press, New York, 1944.

2

SOLUTION OF NONLINEAR EQUATIONS FOR FLOW OF FLUIDS

A DESK calculator is all that's needed to get numerical answers for transient nonlinear flow problems.

Arithmetic and transcendental functions, set down in explicit formulas, eliminate the need to transform the discrete calculus to the infinitesimal calculus used in present-day computer programs to simulate reservoir behavior.

While this chapter deals with the flow of gas in a porous medium —the single phase selected to avoid dealing with relative permeability and the involved phase behavior of a multiphase system — the simplicity of approach opens the door for application to two and three-phase fluid flow. These applications will be discussed.

Linearization of the difficult differential equations will be shown in two cases; one for fluid flow into a well from an infinite reservoir and the other for a limited reservoir. Changes in gas compressibility are accounted for.

Methods presented here are further applied to a field of irregular shape, to a well located in the center of a field, and to a well in an off-center position.

Background. In the first article[1] published on transient fluid flow it was stated that we are dealing with a weight balance of fluids in an infinitesimal element of a reservoir, such that the increase of the weight flux from one surface to its opposite surface represents the loss or depletion of fluid within this element of sand.

This is the equation of continuity given in Lamb[2] and is expressed as

$$\frac{\partial(\rho u)}{\partial x}+\frac{\partial(\rho v)}{\partial y}= - \phi(1-S_{w}) \frac{\partial \rho}{\partial t} \tag{2-1}$$

where ρ is the density of the fluid, and u and v are the corresponding velocities along the respective coordinates, with t, time.

Equation 2-1 associated with Darcy's law for fluid flow, yields

$$\frac{\partial\left(\rho\frac{\partial p}{\partial x}\right)}{\partial x}+\frac{\partial\left(\rho\frac{\partial p}{\partial y}\right)}{\partial y}=\frac{\mu\phi(1-S_w)}{k}\frac{\partial\rho}{\partial t} \tag{2-2}$$

where permeability, k, and viscosity, μ, are accepted as constant.

In that early work[1] another equation was introduced that will have significant application; namely, the density for an expanding fluid is expressed as

$$\rho=\rho_i e^{-c(p_i-p)} \tag{2-3}$$

where c is the compressibility of the fluid. In its application

$$\frac{\partial\rho}{\partial x}=c\rho\frac{\partial p}{\partial x}\ ;\ and\ \frac{\partial\rho}{\partial y}=c\rho\frac{\partial}{\partial y} \tag{2-4}$$

and it follows that Equation 2-2 becomes

$$\frac{\partial^2\rho}{\partial x^2}+\frac{\partial^2\rho}{\partial y^2}=\frac{c\mu\phi(1-Sw)}{k}\frac{\partial p}{\partial t} \tag{2-5}$$

or expressed radially as used in the industry

$$\frac{\partial^2\rho}{\partial r^2}+\frac{1}{r}\frac{\partial\rho}{\partial r}=\frac{\partial\rho}{\partial t_D} \tag{2-6}$$

with

$$t_D=kt/\phi\mu c(1-S_w) \tag{2-7}$$

Equation 2-6 places no limits upon compressiblity, it can be large or small, and there is no restriction as stated in the literature that c has to be small.

However, if it is recognized that c is small, then the expansion for Equation 2-3 becomes

$$\rho\approx\rho_i[1-c(p_i-p)] \tag{2-8}$$

Combining with Equation 2-6 gives

$$\frac{\partial^2 p}{\partial r^2} + \frac{1}{r}\frac{\partial p}{\partial r} = \frac{\partial p}{\partial t_D} \tag{2-9}$$

This equation has been used in water-drive studies developed by the author,[3] and of recent date has received wide application in the interpretation of bottom-hole pressure performances for producing wells.[4]

The solution of these equations as a point source producing from an infinite reservoir was developed by Lord Kelvin of England in the Nineteenth Century. This is given in the text by Carslaw.[5]

This has been applied,[4] expressed by the *Ei* function that conforms to Equation 2-9 to a well producing at a constant rate.

With reference to Equation 2-6 in which density is involved, its identity with the *Ei* function is not reported in the current literature.

Its development is the same as given by Carslaw[5] using the relationship of Equation 2-3, for which

$$\delta \rho = c\rho\delta p \tag{2-10}$$

This expressed in terms of the *Ei* function, yields

$$\rho_i - \rho = \frac{\rho_i q_i c_{gi} \mu}{4\pi kh}\left\{-Ei(-\phi(1-S_w)\mu c_{gi} r^2 / 4kt)\right\} \tag{2-11}$$

used extensively in this chapter, for which the compressiblity refers to gas, although its application is for any fluid conforming to the same boundary conditions.

In treating Equation 2-9, in terms of pressure, the coefficient in Equation 2-11 is $qi\mu/4\pi kh$, given in the literature[4].

Development. In dealing with gases and using the equation of state for gas as density varies directly with pressure, the resulting flow equation in the reservoir, is

$$\frac{\partial^2 p^2}{\partial r^2} + \frac{1}{r}\frac{\partial p^2}{\partial r} = \frac{2\mu\phi(1-S_w)}{k}\frac{\partial p}{\partial t} \tag{2-12}$$

This is reported by Katz,[6] and is observed in Equation 2-2.

This nonlinear differential flow equation is difficult to handle yet the industry, and particularly the gas industry, has had to contend with this for years.

Suppose, however, that the equation of state for gas was absorbed in Equation 2-3 as a changing gas compressibility, then Equation 2-6 would apply. This would be for increments of pressure change within the reservoir where the value for c_g in Equation 2-7 would change for every lowered pressure; yet within that interval of pressure, the identity of Equation 2-6 would remain intact.

This is part of the problem. The second part is how to relate these different diffusivity equations one to another, invoking the changes in compressibilities, as the gas moves through the reservoir to the well bore to satisfy the equation of state that is mandatory in its solution.

This is the series problem developed by the author[7] in which the change in the diffusivity constant is this c_g, the gas compressibility, in place of the formation permeability discussed in the manuscript, although all the physical parameters can undergo a change in traversing to the well bore.

The treatment is the Laplace transformation for an infinite reservoir with the well the point source, but instead of the uniform sand conditions used by Lord Kelvin, the sand and fluid characteristics vary with the distances from the well.

The flow in an infinite reservoir will be considered first, then we shall present this application for an enclosed reservoir. The basic mathematics for the latter[8,9] are given in the literature dealing with a uniform sand, for which the transformation will be made in treating with nonlinear flow.

Infinite reservoir. The series problem now restated in terms of density, compatible with Equation 2-11, is expressed as

$$\rho_i - \rho = \frac{q_i \mu}{2\pi kh} \times \frac{\rho_i}{2} \left\{ c_j \left[-Ei\left(-\frac{c_j \phi \mu (1 - S_w) r^2}{4kt} \right) \right. \right.$$

$$\left. + Ei\left(-\frac{c_j \phi \mu (1 - S_w) r^2{}_j}{4kt} \right) \right] + c_{j+1} \left[-Ei\left(-\frac{c_{j+1} \phi \mu (1 - S_w) r^2{}_j}{4kt} \right) \right.$$

$$+ Ei\left(-\frac{c_{j+1}\phi\mu(1-S_w)r^2{}_{j+1}}{4kt}\right)\Bigg]$$

$$+ \ldots + c_i\left[-Ei\left(-\frac{c_i\phi\mu(1-S_w)r^2{}_n}{4kt}\right)\right]\Bigg\} \tag{2-13}$$

Where $r \leqq r_j$, and $\rho \leqq \rho_j$.

The sweep is from the furthest drainage point in the infinite medium, building up successively with the lowering in pressure as the fluid moves toward the well.

The compressibility c_j refers to gas compressibility for the increments of pressure changes transcribed; with r_n the distance of the first pressure lowering encountered from the initial conditions in the reservoir.

This is illustrated in Fig. 2-1, with each increment of pressure lowering satisfying Equation 2-6 in Equation 2-13.

At the end of each increment, the pressure and its corresonding density fulfill the equation of state for a gas.

The relation for compressiblity is given by

$$c_{j+1} = \frac{-\ln p_j/p_{j+1}}{(p_{j+1}-p_j)} \tag{2-14}$$

that follows from Equation 2-3.

If we take one term in this series of pressure increments in Equation 2-14, this simplifies the nomenclature as well as the facility that this equation has been used. Thus,

$$(\rho_j-\rho)\Big/\frac{q_i\mu\rho_i c_j}{2\pi kh} = \frac{1}{2}\left[-Ei\left(\frac{-c_j\phi\mu(1-S_w)r^2}{4kt}\right)\right.$$

$$\left.+ Ei\left(\frac{-c_j\phi\mu(1-S_w)r^2{}_j}{4kt}\right)\right] \tag{2-15}$$

The left side can take several forms that sustain the gas laws. In terms of pressure, where density varies directly proportional to its pressure, then

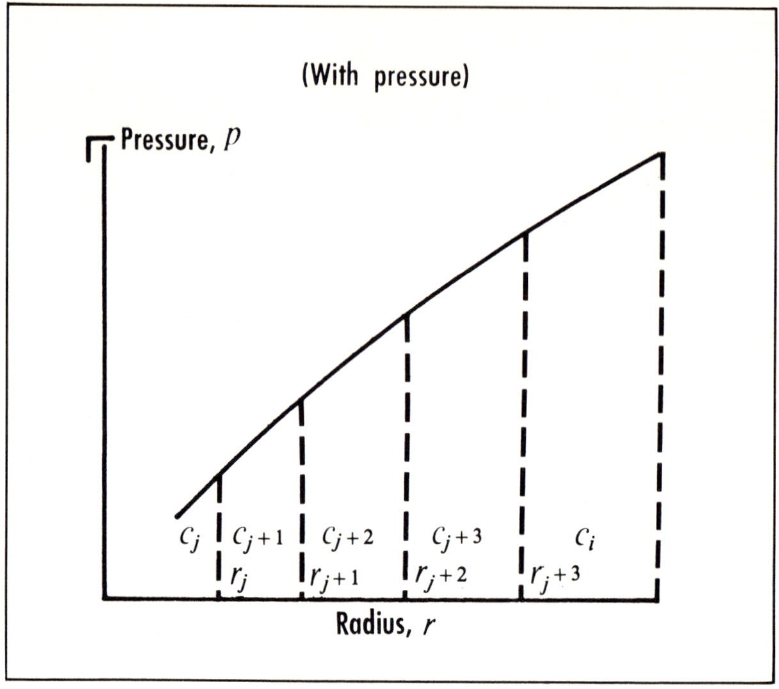

Fig. 2-1 Graphical illustration for changing compressibilities with pressure lowering.

$$(\rho_j - \rho) \Big/ \frac{q_i \mu \rho_i c_j}{2\pi kh} = (p_j - p) \Big/ \frac{q_i \mu p_i c_j}{2\pi kh} \tag{2-16}$$

If we wish to invoke the density change as an integral applying Equation 2-10, the resulting equation is

$$(\rho_j - \rho) \Big/ \frac{q_i \mu \rho_i c_j}{2 6kh} = (p_j^2 - p^2) \Big/ \frac{q_i \mu p_i}{\pi kh} \tag{2-17}$$

Should the gas laws deviate to incorporate a correction factor for the gas, then pressure in Equation 2-16 is the relation p/z_j, as would also apply to Equation 2-14 in making the corrections for densities.

In its application, Equation 2-16 is most adaptable, and an illustrative problem later will show the close comparison between it and Equation 2-17.

In the application of Equation 2-15, radius for its corresponding time is not the independent variable to calculate the pressure within the formation.

This has been customary in the past but density with its corresponding pressure is the independent variable expressed by Equation 2-16 or 2-17 to calculate radius as expressed in the argument of the *Ei* function.

These functions are given in tables[10] readily accessible to the reader.

Further, when the range of a pressure increment is exceeded, a new gas-compressibility factor c_j comes into existence, until the well bore is reached.

Techniques compared. A paper by Bruce et al.[11] gives the solution of unsteady-state gas flow in porous media that forms the basis for the comparison of the work in this chapter. There is a departure in symbols used by the authors and those recommended by AIME. This the reader should recognize and we shall make this transition as readable as possible.

Fig. 5a was chosen from their paper[11] as the nearest fascimile to an infinite reservoir for gas flowing rapidly to a well. This is shown as Fig. 2-2, and includes data from this article.

It is recognized that Bruce[11] treated gas volume as molal volume, whereas our interest is to express these volumes at some measured pressure.

Further, we have little interest in making conversions to engineering units. Any engineer can select whatever units he wishes to use. Such treatment usually disguises the mathematics and makes the presentation more difficult to follow. Here we shall use the interpretation of Darcy's law in its absolute meaning for permeability expressed in c-g-s units.

Thus the symbols from Bruce et al.,[11] converted to AIME nomenclature, are

$$Q = \frac{2q_i\mu}{\pi khp_i} \tag{2-18}$$

and

$$\Theta = \frac{kp_i t}{2\phi\mu r_b^2} \tag{2-19}$$

The Q in this case as illustrated in Fig. 2-2, is $Q = 0.5$, for the infinite reservoir.

The r_b has reference to an enclosed boundary that the authors[11] chose to invoke for their presentation; although it has no specific meaning in Fig. 2-2, it must be identified in their nomenclature. In

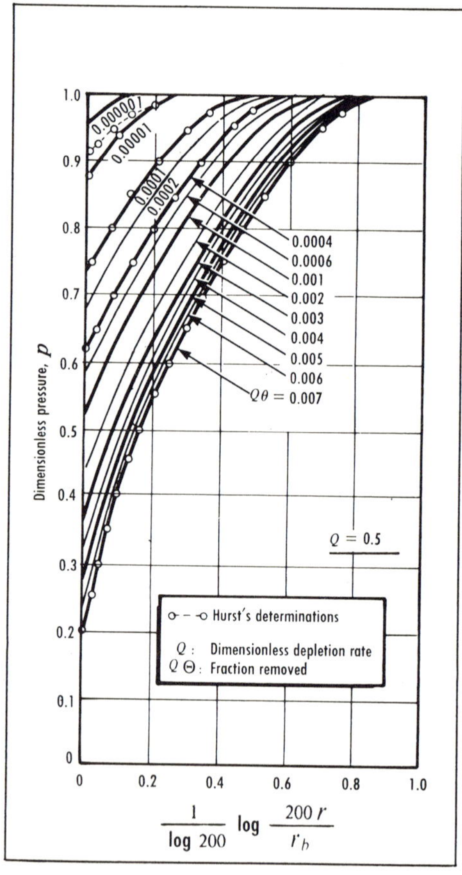

Fig. 2-2 Infinite case, illustration of the non-linear flow.

Fig. 2-2, the ratio of this distant boundary to the inner or well boundary, where gas is produced at a constant rate, is 200:1.

Dimensionless pressure, identified in Fig. 2-2, is the ratio of any lowered reservoir pressure to the initial pressure. Its significance suggests that it have a status symbol, and here it is expressed as

$$\xi = p/p_i \tag{2-20}$$

Thus Equations 2-16 and 2-17 become

$$(p_j - p)\,\frac{q_i \mu p_i c_j}{2\pi kh} = (\xi_j - \xi)\,\frac{q_i \mu c_j}{2\pi kh} \tag{2-21}$$

and

$$(p^2_j - p^2)\,\frac{q_i \mu p_i}{\pi kh} = (\xi^2_j - \xi^2)\,\frac{q_i \mu}{\pi khp_i} \tag{2-22}$$

The gas compressibility given by Equation 2-14 is the relation

$$p_i c_{j+1} = -\;\frac{In\xi_j/\xi_{j+1}}{(\xi_{j+1} - \xi_j)} \tag{2-23}$$

From the definitions given in Equations 2-18 and 2-19, the reader will recognize that the fixed Q in Fig. 2-2 is the same as Bruce's[11] $Q/4 = qi\ \mu/2\ \pi\ khp_i$, as would apply to the formulas developed in this article, recognizing that the *Ei* function in turn has still associated with it one-half value, Equation 2-15. This particularly applies to Equation 2-22.

Dimensionless time as given in the argument of the *Ei* function, Equation 2-15, is expressed as $p_i c_j r^2/8\ \Theta r_b^2 = c_j \phi\ \mu\ r^2/4k\tau$, the $(1 - Sw)$ is inferred in Equation 2-19, although not stated by the earlier authors. Therefore, referring to Fig. 2-2 for a fixed Q, each time curve can be established that applies to a given pressure curve.

This makes the terms used by Bruce et al.[11] in their programming adaptable to the present work for the purpose of checking a changing gas compressibility with changing pressures as now developed to establish radius with time.

The results shown in Fig. 2-2 are a close agreement between the

Two pressure lowerings are enough to indicate how all the subsequent pressure calculations are made relating to radius and time in these nonlinear flow problems.

Recalling Equation 2-18

$$\frac{Q}{2} = \frac{q_i \mu}{\pi k h p_i} = \frac{0.500}{2} = 0.250 \tag{2-24}$$

and

$$\frac{p_i c_j Q r^2}{8 Q \Theta r_b^2} = \frac{\phi(1-S_{w})\mu c_j r^2}{4kt} \tag{2-25}$$

which is the argument of the *Ei* function for a given time, then

$$\frac{p_i c_j Q r^2}{8 Q \Theta r_b^2} = \frac{p_i c_j 0.5 r^2}{8(0.007)4(10)^4}$$

where $r_b = 200$, and

$$\frac{\phi(1-S_{w})\mu c_j r^2}{4kt} = 2.23214(10)^{-4} p_i c_j r^2 \tag{2-26}$$

To use the pressure squared dimensional relationship for a gas, given by Equation 2-22, with the first pressure lowering from $\xi = 1.00$ to $\xi = 0.975$ in the infinite medium, then by Equation 2-15, we have

$$(1-0.975^2)/0.250 = \frac{1}{2}\ [-Ei(-2.25937(10)^{-4} r^2)]$$

where $p_i c_j = 1.01220$ for this first pressure lowering, Table 2-1.

Therefore,

$$-Ei(-2.25937(10)^{-4} r^2) = 0.39500$$

and

$$2.25937(10)^{-4} r^2 = 0.6711 \tag{2-27}$$

which follows from the inverse for the *Ei* function read from tables[10] yield $r = 54.4974$.

Expressed in terms of the abscissa for Fig. 2-2 or log 54.4974/log 200 = 0.75460, shows this point plotted on the graph for this pressure change.

To continue to the next pressure lowering from $\xi = 0.975$ to $\xi = 0.950$, we again use Equation 2-15, interchanging the gas compressibility (Table 2-1).

Thus

$$0.975^2 - 0.950^2/0.250 = \frac{1}{2}\left[-Ei(-2.23214(10)^{-4}(1.03892)r^2) + Ei\left(\frac{-0.6711 \times 1.03892}{1.01220}\right)\right] \quad (2\text{-}28)$$

then $0.38500 = -Ei(-2.31901(10)^{-4}r^2) + Ei(-0.6888)$

and $-Ei(-2.31901(10)^{-4}r^2) = 0.76882$

with $2.31901(10)^{-4}r^2 = 0.3640$

therefore $r = 39.6187$

Thus its abscissa reads as log 39.6187/log 200 = 0.69442 and it is the next point plotted for $\xi = 0.950$.

By this procedure the pressure gradient shown in Fig. 2-2 is reproduced.

Limited reservoir. The solution for the nonlinear differential flow equations results from the author's earlier publication[7] of the point-source solution for a changing permeability in the formation. However, instead of considering permeability by itself, the difference formula so expressed in that work and reproduced in Equation 2-15, is equally applicable to account for the phase relationship and the nonlinearity of these formulas as expressed through a changing gas compressibility with the change in reservoir pressure.

We previously emphasized the *Ei* function because of its shorthand notation and its facility to apply that lends itself to observe reservoir performance.

It will be mentioned later that $P(r, t_D)$ functions of an earlier publication[9] replace the *Ei* function under certain conditions; but its application and usage has the same meaning as the *Ei* function discussed here.

For the limited reservoir, we shall use the transcendental functions published in the literature.[8,9] However, in a later portion we shall show how the *Ei* functions equally apply to reproducing this type of problem.

In the present problem, only the essential parts of Muskat's[8] solution and this writer's work[9] will be incorporated in treating with a well producing at a constant rate in a closed reservoir.

Muskat dealt with a point-source solution, and this writer's work with the well expressed for a fixed well radius.

The nonlinear flow equation applicable to a gas, expressed as a density relationship compatible with what has already been presented in Equation 2-11, is given as

$$\rho_i - \rho = \frac{\rho_i q_i c_{gi} \mu}{2\pi kh} \left\{ \frac{2}{r_b^2} \left(\frac{r^2}{4} + t_D \right) + \ln \frac{r_b}{r} - \frac{3}{4} \right\} \tag{2-29}$$

The time, given by t_D, is expressed by Equation 2-7. The well radius squared, r_w^2, that applies in the denominator, is here referred to as unity in compliance with the work of Bruce et al.[11] in their computer programming.

The results that now follow are illustrated in Fig. 2-3 for Q = 0.05. The circles describe the work of this paper superimposed upon the early author's results.

The immediate interest is the lowering of the reservoir pressure at the field's boundary, r_b, for the gas depleted in time $Q\Theta$.

This is the difference formula mentioned earlier, but instead of areal mapping, the radius is held fixed, and the changing pressures at this boundary determine $Q\Theta$. Thus t_D, in Equation 2-29, is the dependent variable comparable to what has been discussed for the infinite case when radius was the dependent variable and time was held constant.

For the first pressure lowering at this boundary, namely, $\xi = 1$ to $\xi = 0.975$, applying Equation 2-22 in Equation 2-29, for its respective gas compressibility, then

$$(1-\xi^2) / \frac{q_i \mu}{\pi kh p_i} = (1-0.975^2)/0.025 = \frac{2}{200^2} \left(\frac{200^2}{4} + t_D \right) - \frac{3}{4} \tag{2-30}$$

and $t_D = 4.450(10)^4$

From Equation 2-18,

$$Q/2 = \frac{q_i \mu}{\pi k p_i} = 0.05/2 = 0.025$$

for this rate and

$$t_D = \frac{2Q\Theta r_b^2}{Qp_i c_j} = \frac{2Q\Theta 4(10)^4}{0.05 p_i c_j} = \frac{16(10)^5 Q\Theta}{p_i c_j} = \frac{kt}{\phi \mu c_j (1 - S_w)} \qquad (2\text{-}31)$$

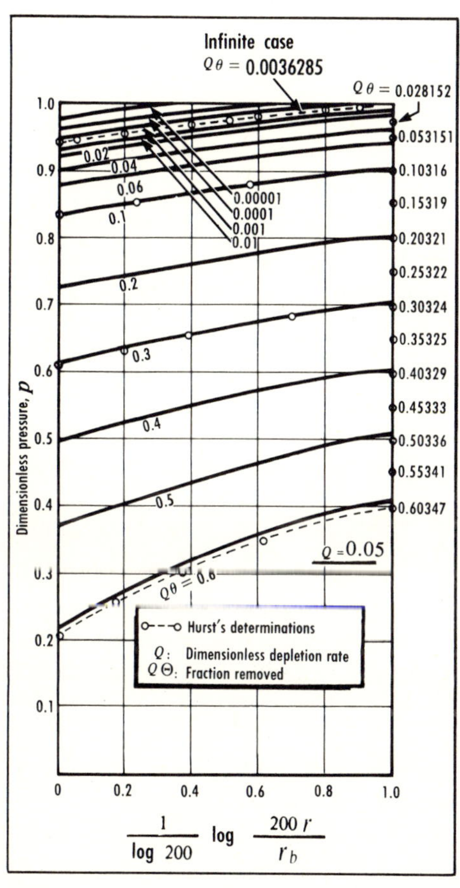

Fig. 2-3 Limited reservoir, illustration of the non-linear flow.

Applying $p_i c_j = 1.01220$ in Equation 2-31 for this first pressure lowering, Table 2-1 gives $Q\Theta = 0.028152$ as shown on the plot.

For the remaining pressure changes given in Table 2-1, the relationship for Equation 2-29 follows in order; namely, the reference to this exterior radius is eliminated, and what we are now dealing with is time only.

Therefore,

$$(\xi_{j+1} - \xi^2_j) \Bigg/ \frac{q_i \mu}{\pi k h p_i} = \frac{2}{r^2_b} (t_{Dj} - t_{D(j+1)}) \tag{2-32}$$

where t_{Dj}, in Equation 2-31, has reference to the existing pressure lowering and c_{j+1}, with $t_{D(j+1)}$, referring to the time of the previous pressure change, but now incorporating the ratio of c_{j+2}/c_{j+1} that converts this to the compressibility prevailing.

The t_{Dj}, by virtue of Equation 2-31, yields the $Q\Theta$ shown plotted at this boundary on the graph (Fig. 2-3).

The areal mapping for the pressure changes within the formation to reach the well bore is identical with what has already been discussed for the infinite case; but here the starting point is for the lowered pressure at the enclosed radius, r_b, for its respective time, $Q\,\Theta$.

This expressed by Equations 2-29 and 2-21 gives

$$(\xi_j - \xi) \Bigg/ \frac{q_i \mu c_j}{2\pi k h} = (\frac{-1}{2r^2_b}(r^2_j - r^2) + In\ \frac{r_j}{r} \tag{2-33}$$

Since time is eliminated, the c_j for the prevailing pressure change is now subscribed by the argument that appears on the left side of Equation 2-33.

These results are likewise illustrated in Fig. 2-3.

For the infinite case, before reaching the enclosed reservoir boundary, this follows identically with what has been stated for the previous case.

The *Ei* function is substituted for the more-involved Bessel function used by both authors[8 9] in the past.

However, the time to reach this external boundary is expressed by the equation for drainage radius when the fluid within the reservoir just begins to flow toward the well bore.

This is given by the relationship[12 13]

$$r_d \cong 2.6408 \sqrt{kt/\phi \mu c_i (1 - S_w)} \quad (2\text{-}34)$$

This is the ratio for distance to the inner well boundary, or the ratio of 200:1 given for this illustrative example, that defines the argument under the square root. Further, for the gas compressibility reported in Table 2-1 at initial conditions, and Equation 2-31, $Q\theta = 0.0036285$ when this boundary is first reached.

The resulting calculations illustrated in Fig. 2-3 for this time follow exactly as in the infinite case.

OTHER APPLICATIONS

Though the early presentation[7] for Equation 2-13 had reference to the changes in permeability within the formation, it is now shown how this also applies to the changes in fluid compressibility.

These variables are not considered by themselves but can undergo simultaneous changes with one another, that include the variation of the net sand thickness within the formation, as well as the other factors associated with Equation 2-7.

There are certain aspects relating to Equation 2-15 that we will now discuss.

What has been shown is a single well producing in a reservoir, but this is also applicable to the study of interference between wells producing in the same formation.

This is expressed in the *Ei* functions by the distances from the interferring wells to the point under consideration, taking into account the pressure lowering and the compressibilities that apply, which are given in Table 2-1.

This equation also lends itself to the application of the superposition theorem when the wells are subject to variable rates.

In this presentation, time, t, by itself has been uneffected. It has reference to the inception of production.

Then the superposition theorem applies and takes into account all varaitions that occur in each term in Equation 2-15. Once this is established we introduce into the time arguments the gas compressibility for its pressure lowering.

With respect to the areal mapping demonstrated in Fig. 2-2, the detailed calculations for the *Ei* functions are not necessary for every $Q\Theta$ curve.

It has been learned early in these calculations this is repetitious—the same steps are performed for each pressure lowering. For a given pressure change the arguments with the *Ei* function are directly related to r^2/t; or expressed in terms of Bruce's work,[11] in going from one $Q\Theta$ curve to another for the same ξ, r^2 is directly proportional to $Q\Theta$.

This is how the $Q\Theta = 0.0001$ curve was developed, the method permits a rapid traverse for each pressure curve for the infinite case.

With reference to $Q\Theta$=0.00001 on this plot, the deviation as shown by the dashed curve is that the *Ei* function does not apply for the time $t_D = 1.46$ that refers to its dimensionless time at the well bore.

This has been mentioned in an earlier presentation[9]; namely, that the point source and the cylindrical solutions are identical only when $t_D \geqslant 100$, although this time range can be much less.

The solution of this problem has been resolved by employing

$$(\xi_{j+1} - \xi_j) \Bigg/ \frac{q_i \mu c_{j+1}}{2\pi kh} = P(r,t_{Dj}) - P(r,t_{D(j+1)}) \tag{2-35}$$

The manner of its solution has been by reconstructing the $P(r, t_D)$ functions from an earlier paper,[1] Fig. 6 in that publication. These are the results that now conform to Bruce's work.[11]

This is one of the omissions in the literature previously mentioned. Although the formulas for $P(r, t_D)$ are given in an earlier paper,[9] the published numerical results are fragmentary.

The dashed curve of Fig. 2-2 shows that the rate $Q = 0.5$ is large in that illustrated problem to reveal a pressure drop for a small time.

Practical technique. The practicality for the use of the *Ei* function is indicated in Equation 2-7. For any oil or gas well, the substitution of the numerical values for the parameters that apply in this formula will show that in a matter of seconds the t_D for a well exceeds 100. Further, since the *Ei* function is an exact solution of Equation 2-6, then in a matter of seconds in absolute time the performance for a well will be transmitted to the reservoir.

In treating with the infinite case and Equation 2-15, it has been mentioned that the sweep is from the most distant point in the reservoir, building up successively with the lowering in pressure until the well bore is reached.

This is dictated by the nature of the problem; however, it has been learned this traverse can take place for a fixed radius to express ξ, with respect to time, Θ.

Figure 2-4 shows this change in well pressure with time. The two sets of data are reported—one from Bruce et al.,[11] and one developed in this chapter.

This is the application of Equation 2-35, but here the $P(r, t_D)$ functions have reference to the $P(t_D)$ functions of an earlier paper,[9] with the well producing at a constant rate in an infinite reservoir.

The *Ei* functions have likewise been treated in this manner, and what it shows is that the well pressures coincide when $Q\theta \geqslant 0.0001$. This is the inaccuracy in the *Ei* function for early times as mentioned, but when this time range is exceeded, the *Ei* function performs exactly as the $P(t_D)$ function.

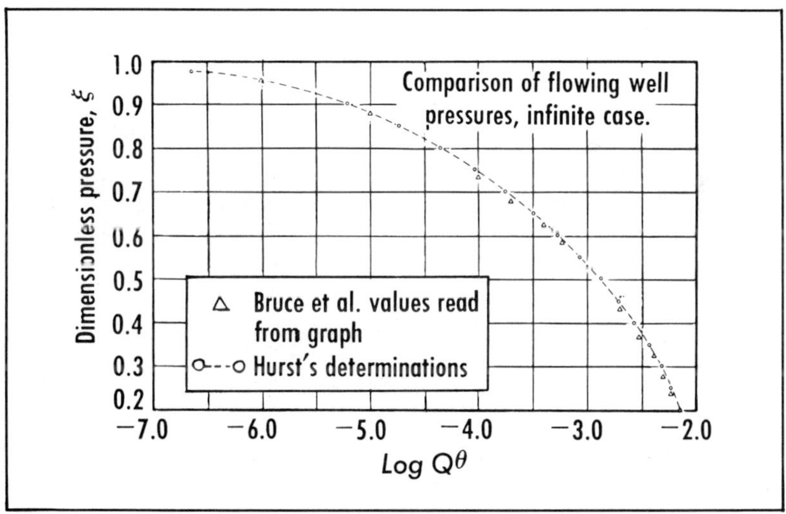

Fig. 2-4 Infinite case, comparison of the flowing well pressures.

With respect to the limited reservoir (Fig. 2-3), this condition for conducting a traverse for a fixed radius, as shown in Fig. 2-4 for the infinite case, no longer applies. All pressure calculations must originate from the enclosed boundary for areal mapping to reach the well pressure as demonstrated in that example.

The controlling factor here is the pressure drop from the enclosed boundary to the well that changes with time because of the change in the lowered reservoir pressure associated with its gas compressibilities.

Thus the distance of a well from an enclosure in the reservoir will have an influence upon the performance of the well.

Compressibility important. Fig. 2-5 invites one's interest. It shows the changing well pressure with time, illustrated in Fig. 2-4, and also shows the calculated bottom-hole pressure when gas compressibility is held fixed, based on initial reservoir conditions. This was the only method we knew to calculate transient fluid flow for gas in the past.

The deviation from the nonlinear fluid-flow formula is evident—for a 50% pressure lowering the error is 25%, and for an 80% pressure lowering, or $\xi = 0.20$, the error is 155%.

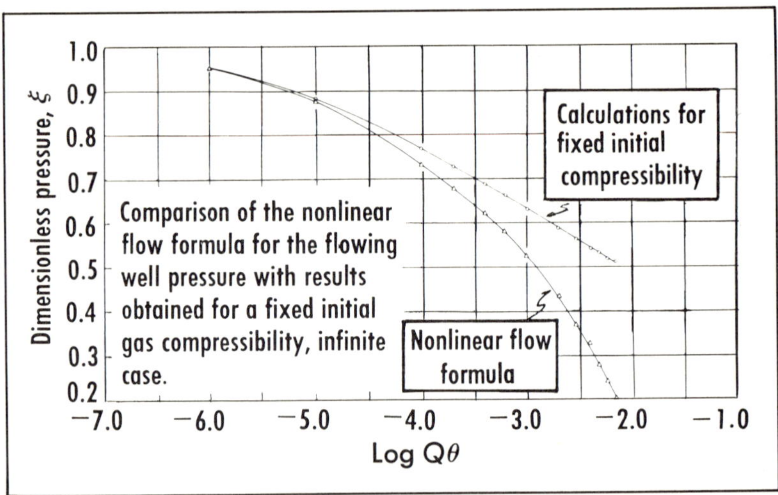

Fig. 2-5 Infinite case, comparison of the Non-linear flow formula for the flowing well pressure with the results obtained for a fixed initial gas compressibility.

For those engaged in gas-deliverability studies, this opens the question as to how much validity can be placed upon the open-flow potentials for gas applied to the entire deliverability study, particularly when these pressures reach the lower levels where such errors can occur.

A major problem is skin effect, particularly when making the sweep to the well bore from the most distant point. Equation 2-22 can account for resistance to gas flow. This is the pressure drop upstream from the skin to reach the lower pressure, which is the flowing well pressure that accounts for this discontinuity.

New solutions—old problems. The methods used in this chapter bring to mind a series of problems not previously solved.

One of these is turbulent flow. This is the condition discussed several years ago when it was reported by others that Darcy's law ceased to apply at the well bore for high rates of gas flow but the pressure drop was directly proportional to the velocity squared. But Darcy's law applies further back in the formation.

Tek et al.,[14] in their presentation on turbulent flow, introduced the thought of matching their formula and making the two solutions coincide as the nonlinear flow formula coming in toward the well subscribes to a logarithm for radius that is almost steady-state flow when the well is being approached. This compares to Tek's further work at some greater distance in the formation.

This matching idea also applies to Muskat's[15] partial-penetrating well producing under steady-state flow, as the same factors are present as described for turbulent flow.

Probably the most important problem of the group is the determination of net sand thickness in a producing well from flow and pressure buildup data. There are areas such as in North Africa, the Bromides in Oklahoma, and the Strawn in West Texas, where electric logs show little character for the interpretation of net sand thickness.

Where we now can account for the nonlinearity in fluid flow, as here described for gas, we have a more-valid means for making reservoir interpretation.

Multiphase flow. During the last few years, much progress has been made in the treatment of two-phase and three-phase fluid flow. Developments in these areas came largely from reconstruction of the performance of old and depleted oil fields and these developments are covered in the next chapter.

The developments entailed the use of simultaneous nonlinear differential flow equations that apply to oil and gas.

Results so far show that what has been obtained for a gas can be applied to the flow of oil and its associated gas released from solution.

Images. Reservoir simulation has already been mentioned in discussing interference between wells, using the nonlinear flow formulas, but here we wish to include the enclosure for a field.

The reproduction of waterflood patterns and gas cycling within rectangles and squares are based on images.

In transient fluid flow the *Ei* functions have been used to represent images as expressed by Mathews et al.,[16] the late Park Jones in his reservoir limit test, and others.

Now, instead of treating with linear boundaries, can this also include boundaries that are not linear but curvatures, comparable to the enclosure for a field?

The procedure is the same as used earlier, but here we are seeking guidelines to include the curvatures.

The concept of an image is fairly simple: If a well is producing in the vicinity of a barrier, its duplicate well, which is its image, placed opposite to the well and across the barrier for the same distance removed, will nullify fluid movement across the barrier.

This is comparable to rotating the well and its included plane 180° through the barrier to reproduce its image. This can apply to any configuration of wells, or the field itself, effected by the same barrier.

Two cases are illustrated that represent enclosures; one with a well located at the center of a circle, and the other with the well off-center from the origin.

The choice for these circular configurations is that the results can be checked.

In offering this presentation, we will depart from the analyses of nonlinear flow that have been discussed, and concentrate entirely on use of the *Ei* functions.

However, once the *Ei* functions are established, its arguments, including the changing gas compressibilities, are applicable to reproducing the nonlinearity of the gas flow, but now conforming to the enclosure for a field.

The first illustrated problem with the well shown at the origin of the circle is given in Fig. 2-6.

Here the delineation of the circle is represented by a series of linear barriers, placed at the enclosure, oriented through equal angles at the origin, with each barrier being tangent to the circle.

The resulting images that apply to these barriers to represent the well form a concentric circle of images to the field.

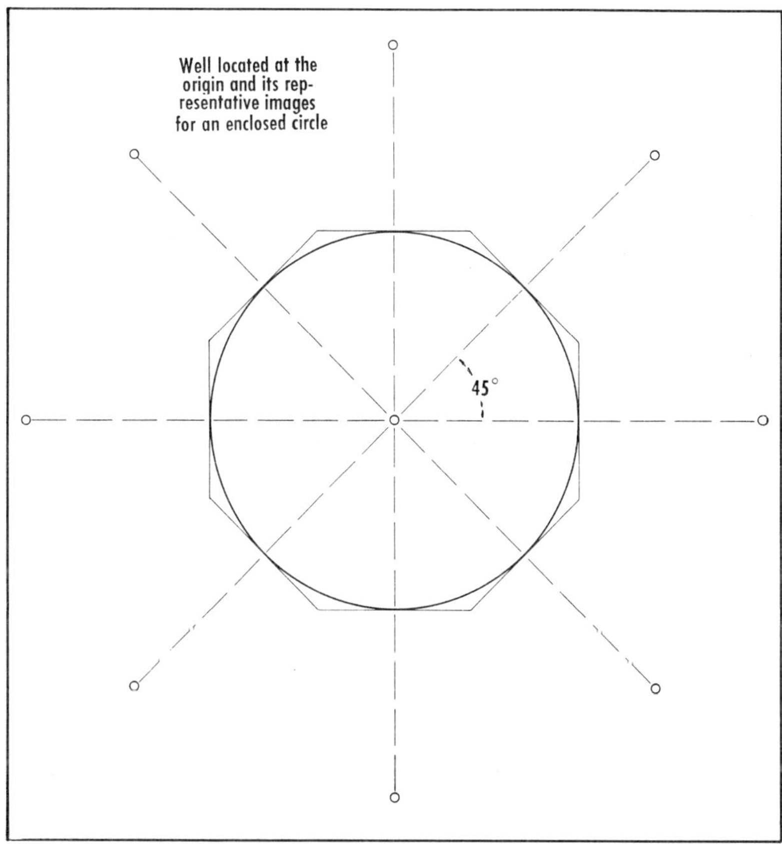

Fig. 2-6 Well Located At The Origin And Its Representative Images For An Enclosed Circle.

prevails. The pattern for these images, although still circular, is not concentric with the field.

This means that cross-flow from one image can invade the territory of another image for continued time, and there is essentially a "cutoff" point in applying the contributions for all the images.

The limiting factor is the slope $2\,\pi/A$, as these images and the well cannot exceed the physical voidage for the well itself.

The means for applying the formula for continued times is still the application of Equation 2-36, but when this slope is reached, the relationship is the straight line as shown in Fig. 2-7.

This is the control exerted for the illustrated problem shown in Fig. 2-9 for the off-center well. The time is $t_D = 15{,}000$, when all the boundaries of the field are effected to give a pressure lowering corresponding to Fig. 2-3 for a fixed time.

The areal mapping shown is the plane through the well and the origin of the circle to give the nonconcentric trough for pressure lowering expected for this type problem.

Two solutions are shown on this plot — one by the method of images as here discussed, and the other, by Muskat[8] for the off-center well.

One has to refer to his publication for the particulars of this solution, for which he uses Green's theorem to simulate the off-center well located within a circle.

His Fourier-Bessel series expansions have no importance here, as

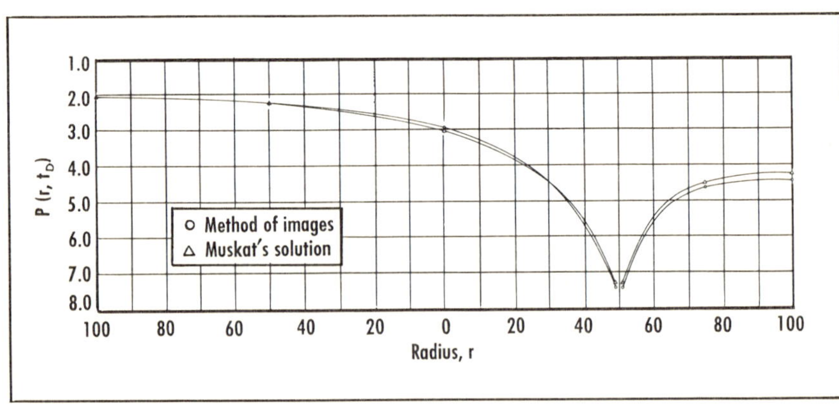

Fig. 2-9 Areal Mapping Of The Off-center Well.

they are nullified with time; but his asymptotic expansions, equivalent to Equation 2-29, are used to check these results.

The close comparison shown on this graph indicates the procedure.

What has been learned here can be restated in treating with a field to include the enclosure. This would correspond to fixing the origin of the system—determined by the center of gravity of the area.

From this point radiates the axes of orientation, separated by equal angles of 45°, but the whole system rotated to give the best coverage for the field.

Where these axes intercept the enclosure, the tangents are drawn to represent the barriers. The images are the rotation of the field 180° through these barriers to represent the particular well involved.

Where a well is located within a wide expanse of the field, one tier of images can suffice; however, where the well is confined within close reaches of the barriers, more tiers will be required by rotating the field through its respective barriers in the vicinity of the well.

The cutoff point in treating with these images is the slope as already defined.

The time this can happen is given by the radius of drainage formula, Equation 2-34.

For the further reaches of the field, this formula can show when transient fluid flow will first take place, to be followed some time later when the whole field will be effected to give the straight-line relationship.

These are the precepts for treating with images; but in constrained or extremely distorted reservoir enclosures, a dual system of origins, or a departure from the number of barriers used, may be needed. This is the individual's own judgment factor.

Conformal mapping. Fig. 2-10 is the artist's conception of conformal mapping applied to a field.

Conformal mapping is a point-to-point transfer of the configuration shown for the field (Fig. 2-10) to an equivalent rectangular system. This has application to steady-state flow.

With respect to waterflooding or gas cycling, this transfer to a rectangle lends itself to the simple treatment if images used by early authors to observe the paths or streamlines from input to producing wells. These in turn can be transferred back to the field proper to develop the fronts encompassed.

To the author's knowledge, this has never been reported in the literature. Nor have we seen references to application of conformal mapping to transient fluid flow or to suitable alternatives.

It is the latter that is of interest.

Conformal mapping can be performed with an electrolytic or a potentiometric model. With respect to a circle, the relationship is

$$p + i\psi = lnr + i\theta = x + iy \tag{2-37}$$

where the pressure and the stream line, ψ, are expressed by polar coordinates for the radius, and the angle of orientation, θ, about the origin.

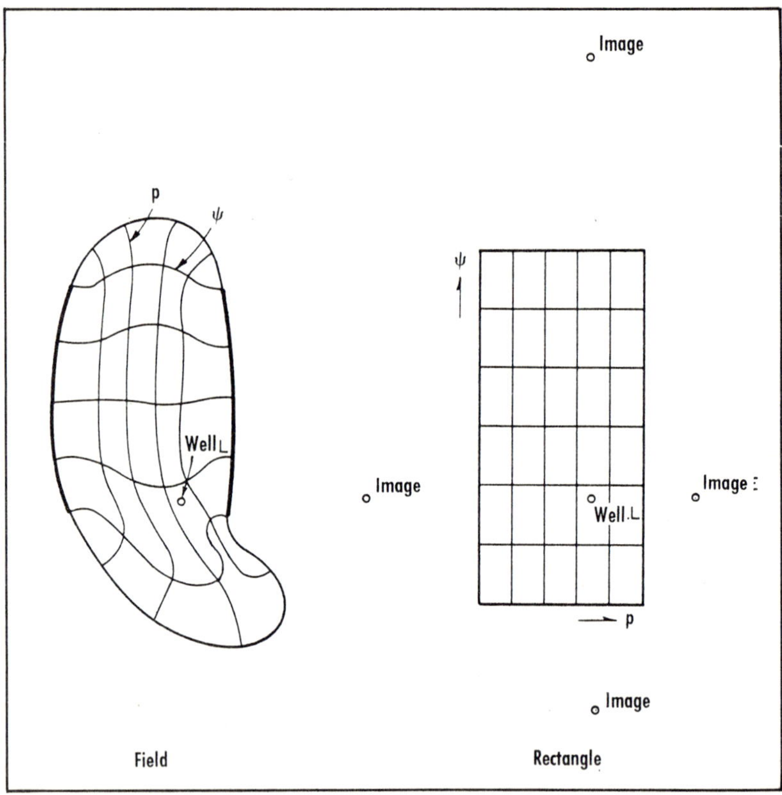

Fig. 2-10 Conformal Mapping Of A Reservoir.

Although this is mathematically correct for what is implied, it has a singularity at the origin that precludes this mapping through the center of the circle.

It is for this reason that the author has undertaken the solution for treating with a circular line source and its counterpart a sink located opposite to it on the perimeter, to accomplish this mapping through the center.

This has now been formularized to learn something of conformal mapping with the solutions for transient fluid flow within a circle.

This is now under investigation, and if it is applicable, not only can this include the enclosure and every inflection exhibited, but an adjoining aquifer as well.

NOMENCLATURE

q —rate of production, L 3/t
ρ —density, m/L^3
p —pressure, m/L t^2
ξ —ratio of lowered to initial pressure, dimensionless
k —permeability, L^2
h —net thickness, L
ϕ —porosity, fraction
μ —viscosity, m/L t
c —compressibility, L t^2/m
S_w —connate water, fraction
r —radius, L
r_d —drainage radius, L
r_b —radius of circular reservoir, L
r_e —radius of external boundary, L
ψ —stream-line velocity, L/t
t —absolute time, t
t_D --dimensionless time, based on unit radius

SUBSCRIPTS

i —initial condition
j —subsequent condition
g —gas

REFERENCES

1. William Hurst, "Unsteady Flow of Fluids in Oil Reservoirs," Physics, Volume 5, January 1934.

2. Sir Horace Lamb, Hydrodynamics, Text, New York Dover Publications, 1945.

3. William Hurst, "Water Influx Into a Reservoir and Its Application to the Equation of Volumetric Balance," Trans. AIME, Vol. 151, 1943.

4. C. S. Matthews and D. G. Russell, "Pressure Build-Up and Flow Tests in Wells," Monograph Volume 1, Henry L. Doherty Series of the AIME, 1967.

5. H. S. Carslaw, Introduction to the Mathematical Theory of the Conduction of Heat in Solids, Text, MacMillian & Co. Ltd., London, 1921.

6. Donald L. Katz, et al., Handbook of Natural Gas Engineering, Text, McGraw-Hill Book Co., New York, 1959.

7. William Hurst, "Interference Between Oil Fields," Trans. AIME, Vol. 219, 1960.

8. Morris Muskat, "The Flow of Compressible Fluids Through Porous Media and Some Problems in Heat Conduction," Physics, Volume 5, March 1934.

9. A. F. van Everdingen and W. Hurst, "The Application of the Laplace Transformation to Flow Problems in Reservoirs," Trans. AIME, Vol. 186, 1949.

10. Tables of Sine, Cosine, and Exponential Integrals, Vols. I and II, Federal Works Agency Work Projects Administration, sponsored by National Bureau of Standards, 1940.

11. G. H. Bruce, D. W. Peaceman, H. H. Rachford, Jr., and J. D. Rice, "Calculations of Unsteady-State Gas Flow Through Porous Media," Trans. AIME, Vol. 198, 1953.

12. William Hurst, Orville K. Haynie, and Richard W. Walker, "New Concepts Extends Pressure Build-Up Analysis," Petroleum Engineer, August 1962: Based on paper, "Some Problems in Pressure Build-Up," SPE 145, by the cited authors, presented at the 1961 Fall Meeting of the AIME, Dallas.

13. William Hurst, "The Radius of Drainage Formula," OGJ, July 14, 1969.

14. M. R. Tek, K. H. Coats and D. L. Katz, "The Effect of Turbulence on the Flow of Natural Gas Through Porous Reservoirs," AIME, Vol. 225, 1962.

15. M. Muskat, Flow of Homogeneous Fluids, Text, J. W. Edwards Inc., 1946.

16. C. S. Matthews, F. Brons and P. Hazebroek, "A Method for Determination of Average Pressure in a Bounded Reservoir," AIME, Vol. 201, 1954.

3

THE SUBSIDIARY EQUATION

THE SUBSIDIARY equation provides a new approach for determining oil saturation in a reservoir at any pressure during the reservoir's history.

This information leads directly to the oil recovery at any period of production, using only a desk calculator.

The subsididary equation is derived from the simultaneous treatment of the basic diffusivity equations for oil and gas released from solution within the reservoir, with both flowing to the well. The fundamental data used are relative permeabilities and PVT data.

Three example fields are covered in this chapter to show results under different formation characteristics. In each case, the oil saturation is shown at various stages as reservoir pressure is lowered.

The significance of the subsidiary equation is that it permits the application of the oil equation alone to determine pressure as a function of radius with time. This is transient fluid flow for two-phase fluid movement.

The gas is accounted for by the balance between the oil and gas produced to equal the oil and gas voided in the pore spaces of the formation.

This approach is an outgrowth of work done for an international oil company. The assignment was to determine the residual oil saturations in place in old and depleted oil fields for which no production records were available.

Subsidiary equation. The subsidiary equation is derived from a mass balance of fluid movement in the reservoir pore space.

This is the diffusivity equation for the voidage of oil in its flow to the well bore joined with the comparable equation for the gas released from solution as formation pressure drops. This gas in turn fills the space voided by the oil.

These complex conditions are related in fairly simple forms of the diffusivity equations by the use of exponential expressions for densities of oil and gas. The pressure gradients that propel the fluids to the well cancel out between these two diffusivity equations, leaving the oil saturation as an explicit function of reservoir pressure.

This results in a nonlinear differential equation between these two variables that is readily solved as illustrated by later examples.

The starting point in these proceedings is the equation of continuity given in Lamb.[1] This has been the basis for the first paper published on transient flow,[2] and applies for multiphase fluid flow.

The equation of continuity can be simply expressed as

$$\frac{\partial(\rho u)}{\partial x} + \frac{\partial(\rho v)}{\partial y} = -\phi(1-S_w)\ \frac{\partial \rho}{\partial t} \qquad (3\text{-}1)$$

where ρ is the density of the fluid and u and v are the corresponding velocities along the respective coordinates, with t, time.

In an earlier paper[2] an exponential relationship was used for the density of the fluid. This simplified the mathematics permitting density, then pressure to be expressed as a function of radius and time.

This procedure is again adopted, so for the gas, this expression is

$$\rho_g = \rho_{gi} e^{-c_{gi}(p_i - p)} \qquad (3\text{-}2)$$

and for the oil, it is

$$\rho_o = \rho_{oi} e^{+c_{oi}(p_i - p)} \qquad (3\text{-}3)$$

Equations 3-2 and 3-3 show that lowering reservoir pressure decreases gas density and increases oil density.

The increase in oil density comes with the release of gas from solution and will be shown graphically later.

Although reference is made to a fixed condition, in actual application the compressibilities, c_o and c_g, are undergoing change with the change in reservoir pressure.

Differentiating Equations 3-2 and 3-3 gives expressions that lend themselves readily to the equation of continuity.

Thus,

$$\frac{\partial \rho_g}{\partial r} = c_g \rho_g \frac{\partial \rho}{\partial r} \tag{3-4}$$

and

$$\frac{\partial \rho_o}{\partial r} = -c_o \rho_o \frac{\partial p}{\partial r} \tag{3-5}$$

Using Equation 3-1, expressed for oil and radial flow, and introducing Darcy's law as it pertains to relative permeability and the flow of oil,

$$u_o = -\frac{k_o}{\mu_o}\frac{\partial p}{\partial r} \tag{3-6}$$

we get

$$\frac{1}{r}\frac{\left(r\rho_o \frac{\partial p}{\partial r}\right)}{\partial r} = \phi \frac{\mu_o}{k_o}\frac{\partial (S_o \rho_o)}{\partial t} \tag{3-7}$$

Equation 3-7 is a weight balance for oil in the reservoir pore spaces. This equation introduces the oil saturation, S_o, associated with time, t, representative of the total pore space that must be accounted for as oil is produced. S_o is the reservoir property we wish to determine in this study.

Therefore,

$$\frac{1}{r}\frac{\partial\left(r \frac{\partial \rho_o}{\partial r}\right)}{\partial r} = -c_o \phi \frac{\mu_o}{k_o}\frac{\partial (S_o \rho_o)}{\partial t} \tag{3-8}$$

and

$$\frac{\partial^2 \rho_o}{\partial r^2} + \frac{1}{r}\frac{\partial \rho_o}{\partial r} = -c_o \phi \frac{\mu_o}{k_o}\frac{\partial (S_o \rho_o)}{\partial t} \tag{3-9}$$

Expressing the left-hand side of Equation 3-9 in terms of pressure, and recalling Equation 3-5,

$$\frac{\partial^2 \rho_o}{\partial r^2} = c_o^2 \rho_o \left(\frac{\partial p}{\partial r}\right)^2 - c_o \rho_o \frac{\partial^2 p}{\partial r^2} \tag{3-10}$$

since c_o is small, Equation 3-9 simplifies to

$$\frac{\partial^2 p}{\partial r^2} + \frac{1}{r}\frac{\partial p}{\partial r} = \frac{\phi}{\rho_o}\frac{\mu_o}{k_o}\frac{\partial (S_o \rho_o)}{\partial t} \tag{3-11}$$

Differentiation of the right-hand side of Equation 3-11 in terms of pressure, yields

$$\frac{\partial (S_o \rho_o)}{\partial t} = \left(-S_o c_o \rho_o + \rho_o \frac{\partial S_o}{\partial p}\right)\frac{\partial p}{\partial t}$$

and

$$\frac{\partial^2 p}{\partial r^2} + \frac{1}{r}\frac{\partial p}{\partial r} = \varphi \frac{\mu_o}{k_o} c_o \left(\frac{1}{c_o}\frac{\partial S_o}{\partial p} - S_o\right)\frac{\partial p}{\partial t} \tag{3-12}$$

In a similar manner, using Darcy's Law for relative permeability and flow of gas

$$u_g = -\frac{k_g}{\mu_g}\frac{\partial p}{\partial r} \tag{3-13}$$

Combined with the equation of continuity as in development of Equation 3-7, we get

$$\frac{1}{r}\frac{\partial \left(r \rho_g \frac{\partial p}{\partial r}\right)}{\partial r} = \varphi \frac{\mu_g}{k_g}\frac{\partial (S_g \rho_g - S_o \gamma_g)}{\partial t} \tag{3-14}$$

where, S_g, is the gas saturation in the interstices, and γ_g is the weight of gas liberated from solution per volume of oil at reservoir conditions determined from PVT analysis.

Since this is gas that is added, and separate from the depletion within the gas space that is being voided of gas, it is negative.

Performing the calculations performed earlier for oil, then

$$\frac{\partial^2 \rho_g}{\partial r^2} + \frac{1}{r}\frac{\partial \rho_g}{\partial r} = c_g \varphi \frac{\mu_g}{k_g}\frac{\partial(S_g \rho_g\ S_o \gamma_g)}{\partial t} \tag{3-15}$$

Expressing the left-hand side of Equation 3-15 in terms of pressure,

$$\frac{\partial^2 \rho_g}{\partial r^2} = \rho_g c^2{}_g \left(\frac{\partial p}{\partial r}\right)^2 + \rho_g c_g \frac{d^2 p}{\partial r^2} \tag{3-16}$$

and recalling Equation 3-4,

$$\frac{\partial^2 \rho}{\partial r^2} + \frac{1}{r}\frac{\partial p}{\partial r} = \frac{\varphi \mu_g}{\rho_g k_g}\frac{\partial(S_g \rho_g - S_o \gamma_g)}{\partial t} \tag{3-17}$$

where c_g is small.

Differentiation of the right of Equation 3-17 in terms of pressure, yields

$$\frac{\partial(S_g p_g - S_o \gamma_g)}{\partial t} = \left(S_g \frac{\partial \rho_g}{\partial p} + \rho_g \frac{\partial S_g}{\partial p} - S_o \frac{\partial \gamma_g}{\partial p} - \gamma_g \frac{\partial S_o}{\partial p}\right)\frac{\partial p}{\partial t}$$

$$= \left(c_g S_g \rho_g + \rho_g \frac{\partial S_g}{\partial p} - S_o \frac{\partial \gamma_g}{\partial p} - \gamma_g \frac{\partial S_o}{\partial p}\right)\frac{\partial p}{\partial t} \tag{3-18}$$

Therefore, what is involved in Equation 3-17 is

$$\frac{\partial^2 p}{\partial r^2} + \frac{1}{r}\frac{\partial p}{\partial r} =$$

$$\phi \frac{\mu_g c_g}{k_g} \times \left(S_g + \frac{1}{c_g}\frac{\partial S_g}{\partial p} - \frac{S_o}{c_g p_g}\frac{\partial \gamma_g}{\partial p} - \frac{\gamma_g}{c_g p_g}\frac{\partial S_o}{\partial p}\right)\frac{\partial p}{\partial t} \tag{3-19}$$

Equating the right-hand sides of Equations 3-12 and 3-19.

$$\phi \frac{\mu_o}{k_o} c_o \left(\frac{1}{c_o} \frac{\partial S_o}{\partial p} - S_o \right) =$$

$$\phi \frac{\mu_g c_g}{k_g} \left(1 - S_w - S_o - \frac{1}{c_g} \frac{\partial S_o}{\partial p} - \frac{S_o}{c_g \rho_g} \frac{\partial \gamma_g}{6p} - \frac{\gamma_g}{c_g \rho_g} \frac{\partial S_o}{\partial p} \right) \quad \text{(3-20)}$$

where $S_g = 1 - S_W - S_o$; and $\partial S_g / \partial p = -\partial S_o / \partial p$.

Further, let

$$f(S_o) = \frac{\mu_o c_o k_g}{\mu_g c_g k_o} \quad \text{(3-21)}$$

This is a function of reservoir pressure and oil saturation as expressed through Equations 3-12 and 3-19 as instantaneous diffusivity equations, representative of the coefficients.

Thus

$$f(S_o) \left(\frac{1}{c_o} \frac{\partial S_o}{\partial p} - S_o \right) =$$

$$\left(1 - S_w - S_o - \frac{1}{c_g} \frac{\partial S_o}{\partial p} - \frac{S_o}{c_g \rho_g} \frac{\partial \gamma_g}{\partial p} - \frac{\gamma_g}{c_g \rho_g} \frac{\partial S_o}{\partial p} \right) \quad \text{(3-22)}$$

Solving for $\partial S_o / \partial p$ that appears in the relation, we have

$$\frac{\partial S_o}{\partial p} = \frac{\left[\left(f(S_o) - 1 - \frac{1}{c_g \rho_g} \frac{\partial \gamma_g}{\partial p} \right) S_o + 1 - S_w \right]}{\frac{1}{c_o} \left[f(S_o) + \frac{c_o}{c_g} + \frac{\gamma_g c_o}{\rho_g c_g} \right]} \quad \text{(3-23)}$$

The integration of this formula from the initial reservoir pressure, P_b, and oil saturation, $1\text{-}S_w$, to a lower pressure condition, yields

$$S_o - 1 + S_w = \int_{P_b}^{p} \frac{\left[\left(f(S_o) - 1 - \frac{1}{c_g \rho_g} \frac{\partial \gamma_g}{\partial p}\right) S_o + 1 - S_w\right] dp}{\frac{1}{c_o}\left[f(S_o) + \frac{c_o}{c_g} + \frac{\gamma_g c_o}{\rho_g c_g}\right]} \tag{3-24}$$

where $p \leqslant p_b$, and S_o is the resulting oil saturation at this lowered pressure. The fact that S_o is also contained within the integrand offers no complication in the solution of Equation 3-24, recognized as a nonlinear differential equation.

The procedure used to determine S_o in this formula is to assume a linear variation between S_o and reservoir pressure, p. The choice for S_o at the lower abandonment pressure is arbitrary.

Using the chosen value of S_o in the integrand of Equation 3-24 gives a calculated value for S_o.

The average of the assumed and calculated S_o values is the basis for the next assumed value for S_o in the formula. This usually requires four repetitive calculations to give a rapid convergence for oil saturation vs. pressure.

Application. Figs. 3-1, 3-2, and 3-3, show oil saturations in place

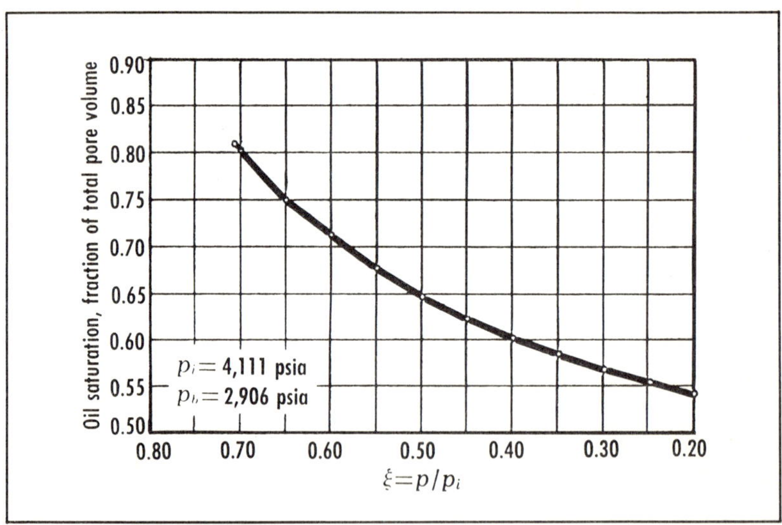

Fig. 3-1 Oil Saturation Versus Pressure Lowering. Muddy Formation, Hilight Area, Campbell County, Wyoming.

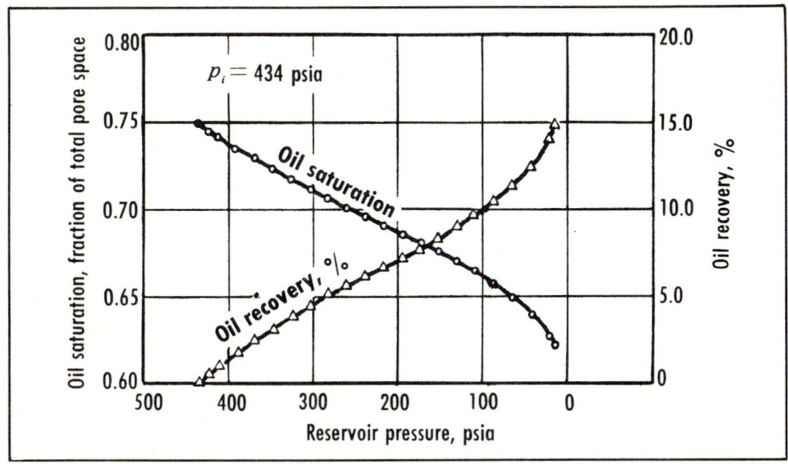

Fig. 3-2 Oil Saturation And Oil Recovery Vs. Reservoir Pressure. Tertiary Formation, Pennsylvania Series, Field A.

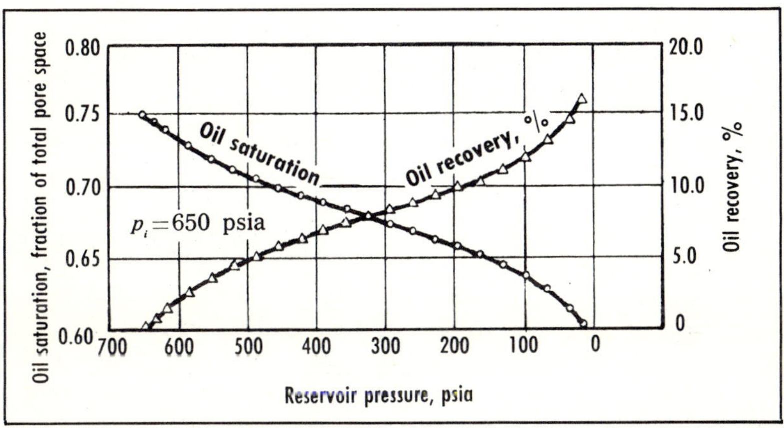

Fig. 3-3 Oil Saturation And Oil Recovery Vs. Reservoir Pressure. Tertiary Formation Mississippian Series, Field B.

vs. pressure for the fields considered in this paper and developed from the mathematics.

The first, the Hilight Area, Campbell County, Wyo., was the test case to determine if the method applied.

This field was chosen because the physical data were at hand, the field is a major producer, and the author had been involved in its early discovery.[3]

The reservoir is the Muddy formation at 9,000 ft, with a connate water content of 19% to give an initial oil saturation of 81% shown on the plot.

This is a typical high-pressure field, undersaturated with gas at the beginning, with a gas-oil ratio of 1,310 cu ft/bbl.

The remaining two cases are developed from postmortem data on Fields A and B. These are old, depleted oil fields produced at the turn of the century for which no information is available.

The procedure was straightforward to acquire the necessary data for this study.

Representative core samples were sent to Core Laboratories Inc. of Dallas for relative permeability determination. Results are shown in Figs. 3-5 and 3-6 for Field A.

The PVT analysis was performed by PVT Laboratores Inc. of Houston.

The instructions to this laboratory were to reconstruct the PVT analysis by charging a representative sample of the crude from the

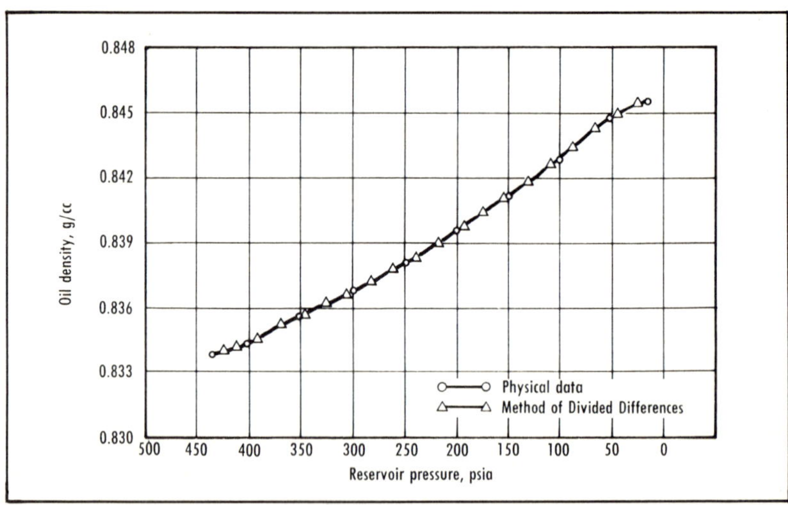

Fig. 3-4 Density Of Saturated Oil At Reservoir.

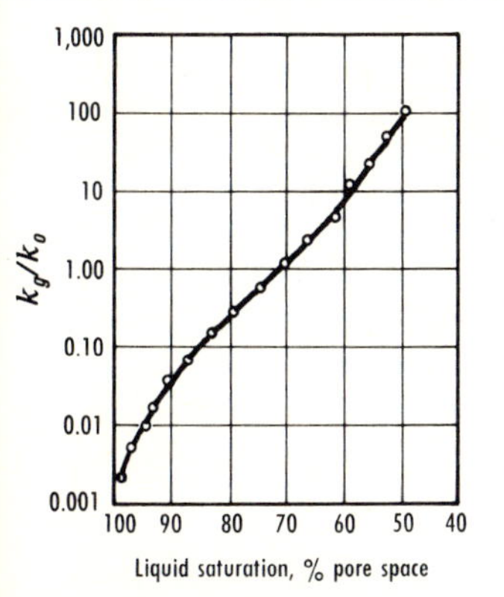

Fig. 3-5 Relative Permeabilities, Ratio Of Gas To Oil, Field A.

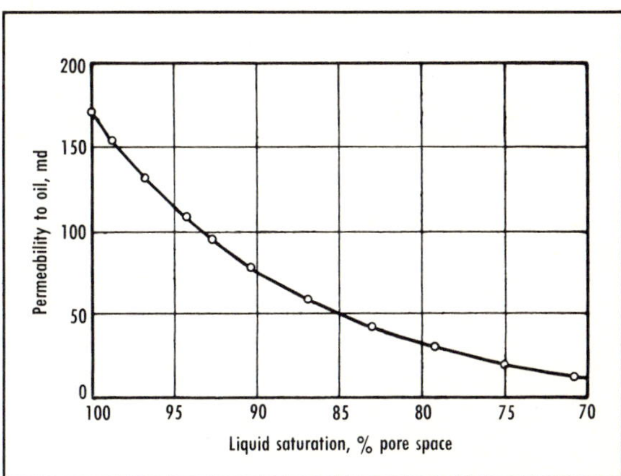

Fig. 3-6 Relative Oil Permeability, Field A.

field with gas to reach an initial saturation pressure of 434 psia for Field A. Some of these data are shown in Fig. 3-4, but most are reported in table form.

For the higher pressure of Field B, the PVT analysis for Field A was extrapolated by methods used in material-balance studies which will be discussed later.

The core analysis and relative permeabilities for Field B are separate from Field A, although not reported in this presentation.

Based on these data, and the application of Equation 3-24, the graphs shown in Figs. 3-2 and 3-3 were developed.

Likewise shown are the recoveries of oil originally in place for each lowering in reservoir pressure. This is expressed by the relation

$$\% \text{ recovery} = \frac{S_{oi}/B_{oi} - S_o/B_o}{S_{oi}/B_{oi}} \times 100 \qquad (3\text{-}25)$$

that accounts for the oil voided by the changes in oil saturation.

Thus for Fig. 3-2 the recovery at atmospheric pressure is 14.9% and for Fig. 3-3 is 16.1%. These are compatible with recoveries reported for analogous fields.

The mechanics of calculating Equation 3-24 show that the PVT data are most significant while relative permeabilities, Fig. 3-5, play a minor role.

Physical data. PVT data are known and have been used by those who have conducted material-balance studies in the past. Therefore, no detailed discussion will be entered upon here except to refer to terms used in the subsidiary equation that are different from the usual application of PVT data.

For a comprehensive understanding of the subject, the reader is referred to an excellent text by Amyx et al.[4]

Two essential phenomena are reported in PVT analyses, flash liberation and differential liberation.

Flash liberation is a measurement of the composite volumes of gas liberated from solution and its associated oil.

Differential liberation refers only to this saturated oil volume at reservoir conditions.

The difference between these two is the volume of gas liberated.

This is the application that has been used to determine γ_g in the subsidiary equation that expresses the weight of gas liberated from solution per unit volume of oil.

With reference to the Hilight field, flash liberation is expressed by the correlation developed by S. E. Buckley.

$$y = (p_b - p)/p\Delta V \tag{3-26}$$

where ΔV is the incremental volume increase of these composite mixtures and referred to a unit volume of oil at saturation pressure.

The differential liberation in turn is expressed by the relationship established by the author, where

$$\log \overline{\Delta V} = a + b\log(p_b - p) \tag{3-27}$$

and $\overline{\Delta V}$ is the decrease in oil volume from saturation pressure, also referred to a unit volume of oil at the bubble-point pressure.

Thus, for the Hilight field, the ratio of $(V - \overline{V})/\overline{V}$ is the volume of gas liberated from solution per unit volume of oil at reservoir conditions.

The gravity and correction factor for the liberated gas are available from the PVT analyses to permit determination of gas density. The product of these two factors is the γ_g term referred to in the subsidiary equation.

These correlations have also been used to extrapolate the PVT data for Field A to the higher pressure of Field B to find its γ_g values.

It is not always necessary to use these correlations. For Field A, density of gas liberated from solution, corrected to reservoir conditions, and the formation volume factor for its associated oil, yielded γ_g.

Illustrations of the correlations mentioned for the Hilight field are given in a paper by Kennedy.[3] Amyx et al.[4] make further mention of these correlations, although neither author has published his results which each considered elementary although of importance in PVT analyses.

What is needed in Equation 3-24 is the rate of the change for each of these physical parameters with respect to pressure. This refers to γ_g and the instantaneous compressibilities for oil and gas obtained from their density relationships described in Equations 3-4 and 3-5, namely

$$c_g = \frac{1}{\rho_g} \frac{\partial \rho_g}{\partial p} \tag{3-28}$$

and

$$c_o = -\frac{1}{\rho_o} \frac{\partial \rho_o}{\partial p} \tag{3-29}$$

To determine these differentials for the Hilight field, empirical curves were developed from the physical data. However, these were too time-consuming and did not yield a realistic interpretation of every inflection that could occur in the data.

Several attempts were made to find an acceptable method to interpolate and differentiate these data, including programming procedures. The search finally narrowed to the Method of Divided Differences as the most appropriate to this application. The reason for this is that the pressure changes in a PVT analysis are not necessarily equal but are those which are most convenient for the laboratory to report.

There is no reference to the Method of Divided Differences in the literature, so an illustrative example for Field A's saturated oil density is shown in Table 3-1. This table also includes a sample calculation for the interpolation and differentiation for the oil compressibility given in Equation 3-29.

TABLE 3-1
Oil density by method of divided differences

p, psia	ρ_o, g/cc	Δ	Δ^2
434	0.8339		
		$-1.47058(10)^{-5}$	
400	0.8344		$+1.34455(10)^{-7}$
		$-2.60000(10)^{-5}$	
350	0.8357		$-0.40000(10)^{-7}$
		$-2.20000(10)^{-5}$	
300	0.8368		$+0.40000(10)^{-7}$
		$-2.60000(10)^{-5}$	
250	0.8381		$+0.40000(10)^{-7}$
		$-3.00000(10)^{-5}$	
200	0.8396		$+0.20000(10)^{-7}$
		$-3.20000(10)^{-5}$	
150	0.8412		$+0.20000(10)^{-7}$
		$-3.40000(10)^{-5}$	
100	0.8429		$+0.40000(10)^{-7}$
		$-3.80000(10)^{-5}$	
50	0.8448		$-1.79803(10)^{-7}$
		$-2.26628(10)^{-5}$	
14.7	0.8456		

Divided Differences

$$\xi = 0.975,\ p = 423.2 \text{ psia}$$

Interpolation,

$$\begin{aligned}\rho_o &= 0.8339 + (423.2 - 434)\,(-1.47058)\,(10)^{-5}\\ &\quad + (423.2 - 434)\,(423.2 - 400)\,(1.34455)\,(10)^{-7}\\ &= 0.83403 \text{ gms/cc}\end{aligned}$$

Differentiation,

$$\begin{aligned}\frac{\partial \rho_o}{\partial p} &= -1.47058(10)^{-5}\\ &\quad + (423.2 - 434)\,(1.34455)\,(10)^{-7}\\ &\quad + (423.2 - 400)\,(1.34455)\,(10)^{-7}\\ &= -1.30385\,(10)^{-5}\end{aligned}$$

Oil Compressibility, Eq. 3-4

$$\begin{aligned}c_o &= \frac{-1\,(-1.30385)\,(10)^{-5}}{0.83403}\\ &= 1.56331(10)^{-5} \text{ vol/vol/psi}\end{aligned}$$

The first two columns of Table 3-1 show the density of the oil vs. reservoir pressure. This is laboratory data from the differential liberation, and it has been requested in all the PVT analyses used in this work.

The third column, expressed by Δ, is the ratio of the change of this oil density with respect to its pressure.

The fourth column, Δ^2, is the rate of change in Δ, with respect to the overall pressure change that has occurred, and corresponds to a second-order differential, although expressed in increments.

Convenient pressure ratios ($\xi = 1.00, 0.975, 0.950, 0.900$, etc.), were selected for use not only in the subsidiary equation but as the range of pressure lowerings involved in the nonlinear differential flow equations for two-phase fluid flow that will follow. The example calculation in Table 3-1 uses $\xi = 0.975$.

The arrows in this table indicate the directions that these interpolations can be used. Thus the pressure interval that ξ refers to determines the values for Δ and Δ^2. For the lower pressure range, the path can be reversed as shown by its arrow.

The results are given in Fig. 3-4, which illustrates the comparison of physical data with the values established by the Method of Divided Differences. It is to be observed that every inflection is faithfully reproduced from the PVT data.

Finally, the summarized data that include the calculated PVT values for Field A and its relative permeabilities are given in Table 3-2. The results are likewise listed, determined from Equation 3-24 to yield the oil saturation in situ versus the reservoir pressure illustrated in Fig. 3-2.

Methods described here will be applied to numerical examples later.

Fluid Flow. Two sources form the background for the two-phase fluid-flow problem

In the first, "Interference Between Oil Fields,"[6] the series problem is presented for the variation in permeability to fluid flow. This is the extension of the Lord Kelvin solution[8] for a point source, but instead of a uniform medium, these variations are incorporated treating with the Laplace Transformations.

The second is the previous chapter which recognizes that fluid characteristics as well as formation properties can vary. These

TABLE 3-2
Physical data for Field A

ξ	p, psia	γ-g, g/cc	$\partial\gamma g/\partial p$, g/cc/psi	ρ_o, g/cc	C_o, vol/vol/psi	ρ_g, g/cc	C_g, vol/vol/psi	μ_o, cp	k_g/k_o	S_o, Oil sat'n	$\partial S_o/\partial p$, Oil sat'n/psi
1.000	434.0	$0.00000(10)^{-3}$	$-1.83584(10)^{-5}$	0.83390	$1.21528(10)^{-5}$	$31.622(10)^{-3}$	$3.71020(10)^{-3}$	5.28000	0	0.75000	$4.35400(10)^{-4}$
0.975	423.2	$0.20163(0)^{-3}$	$-1.89811(10)^{-5}$	0.83403	$1.56331(10)^{-5}$	$30.376(10)^{-3}$	$3.73324(10)^{-3}$	5.29490	0.00070	0.74578	$3.57248(10)^{-4}$
0.950	412.3	$0.41195(10)^{-3}$	$-1.96095(10)^{-5}$	0.83418	$1.91441(10)^{-5}$	$29.162(10)^{-3}$	$3.75279(10)^{-3}$	5.31086	0.00126	0.74207	$3.25502(10)^{-4}$
0.900	390.6	$0.85462(10)^{-3}$	$-2.11511(10)^{-5}$	0.83466	$3.26456(10)^{-5}$	$26.872(10)^{-3}$	$3.77902(10)^{-3}$	5.34617	0.00250	0.73545	$2.88480(10)^{-4}$
0.850	368.9	$1.3228(10)^{-3}$	$-2.19975(10)^{-5}$	0.83523	$3.05449(10)^{-5}$	$24.787(10)^{-3}$	$3.80643(10)^{-3}$	5.38480	0.00325	0.72928	$2.80931(10)^{-4}$
0.800	347.2	$1.8127(10)^{-3}$	$-2.39984(10)^{-5}$	0.83575	$2.41986(10)^{-5}$	$22.803(10)^{-3}$	$3.83782(10)^{-3}$	5.42532	0.00415	0.72319	$2.79949(10)^{-4}$
0.750	325.5	$2.3304(10)^{-3}$	$-2.37146(10)^{-5}$	0.83622	$2.62610(10)^{-5}$	$20.948(10)^{-3}$	$3.98291(10)^{-3}$	5.46775	0.00510	0.71728	$2.63730(10)^{-4}$
0.700	303.8	$2.8420(10)^{-3}$	$-2.34307(10)^{-5}$	0.83671	$2.83204(10)^{-5}$	$19.182(10)^{-3}$	$4.13689(10)^{-3}$	5.51203	0.00615	0.71171	$2.49967(10)^{-4}$
0.650	282.1	$3.3444(10)^{-3}$	$-2.30738(10)^{-5}$	0.83725	$3.03756(10)^{-5}$	$17.522(10)^{-3}$	$4.27165(10)^{-3}$	5.55938	0.00715	0.70636	$2.40911(10)^{-4}$
0.600	260.4	$3.8444(10)^{-3}$	$-2.30132(10)^{-5}$	0.83781	$3.24274(10)^{-5}$	$15.928(10)^{-3}$	$4.52486(10)^{-3}$	5.60713	0.00830	0.70120	$2.35441(10)^{-4}$
0.550	238.7	$4.3431(10)^{-3}$	$-2.29524(10)^{-5}$	0.83843	$3.51276(10)^{-5}$	$14.400(10)^{-3}$	$4.78340(10)^{-3}$	5.65397	0.00960	0.69611	$2.30931(10)^{-4}$
0.500	217.0	$4.8405(10)^{-3}$	$-2.28916(10)^{-5}$	0.83908	$3.61348(10)^{-5}$	$12.929(10)^{-3}$	$5.16335(10)^{-3}$	5.70148	0.0120	0.69124	$2.14204(10)^{-4}$
0.450	195.3	$5.3368(10)^{-3}$	$-2.31976(10)^{-5}$	0.83975	$3.71396(10)^{-5}$	$11.506(10)^{-3}$	$5.57204(10)^{-3}$	5.74999	0.0138	0.68664	$2.13984(10)^{-4}$
0.400	173.6	$5.8737(10)^{-3}$	$-2.62894(10)^{-5}$	0.84043	$3.81424(10)^{-5}$	$10.132(10)^{-3}$	$6.17301(10)^{-3}$	5.79962	0.0157	0.68176	$2.35550(10)^{-4}$
0.350	151.9	$6.4777(10)^{-3}$	$-2.93812(10)^{-5}$	0.84114	$3.91421(10)^{-5}$	$8.7914(10)^{-3}$	$6.93609(10)^{-3}$	5.85492	0.0184	0.67645	$2.53387(10)^{-4}$
0.300	130.2	$7.1778(10)^{-3}$	$-3.29760(10)^{-5}$	0.84185	$3.98931(10)^{-5}$	$7.4961(10)^{-3}$	$7.90010(10)^{-3}$	5.91222	0.0210	0.67061	$2.82123(10)^{-4}$
0.250	108.5	$7.9041(10)^{-3}$	$-3.39691(10)^{-5}$	0.84260	$4.19179(10)^{-5}$	$6.2195(10)^{-3}$	$9.39657(10)^{-3}$	5.98027	0.0242	0.66426	$2.99540(10)^{-4}$
0.200	86.8	$8.6520(10)^{-3}$	$-3.49620(10)^{-5}$	0.84349	$5.00817(10)^{-5}$	$4.9676(10)^{-3}$	$11.5309\ (10)^{-3}$	6.05960	0.0285	0.65758	$3.24267(10)^{-4}$
0.150	65.1	$9.3498(10)^{-3}$	$-3.86463(10)^{-5}$	0.84432	$4.07901(10)^{-5}$	$3.7254(10)^{-3}$	$15.3545\ (10)^{-3}$	6.15026	0.0340	0.64999	$3.75044(10)^{-4}$
0.100	43.4	$10.2632(10)^{-3}$	$-4.55385(10)^{-5}$	0.84498	$3.15231(10)^{-5}$	$2.4850(10)^{-3}$	$22.9867\ (10)^{-3}$	6.28109	0.0415	0.64081	$4.91381(10)^{-4}$
0.050	21.7	$11.326(10)^{-3}$	$-5.24307(10)^{-5}$	0.84548	$2.22749(10)^{-5}$	$1.2463(10)^{-3}$	$45.7698\ (10)^{-3}$	6.58299	0.0540	0.62775	$7.37552(10)^{-4}$
0.33871	14.7	$11.701(10)^{-3}$	$-5.46539(10)^{-5}$	0.84560	$1.92949(10)^{-5}$	$0.84709(10)^{-3}$	$67.3092\ (10)^{-3}$	6.7000	0.0640	0.62147	$9.05890(10)^{-4}$

$\rho_g = \mu_g$ 0.010475 cp, Constant

variations are introduced for the nonlinear flow of gases in a formation as gas compressibilities change with reservoir pressure decline.

It is recognized that this is the dynamic condition for two-phase fluid flow; and instead of treating with two separate diffusivity equations, one equation for oil defines transient fluid flow, and the subsidiary equation is the relation between the oil and gas.

Thus Equation 3-9, that expresses the oil equation in terms of density, can be written as

$$\frac{\partial^2 \rho_o}{\partial r^2} + \frac{1}{r}\frac{\partial \rho_o}{\partial r} = \phi \frac{\mu_o}{k_o} c_o \left(\frac{1}{c_o}\frac{\partial S_o}{\partial p} - S_o \right) \frac{\partial \rho_o}{\partial t} \qquad (3\text{-}30)$$

The relation for oil and gas liberated from solution is given by the dimensionless time in

$$t_D = \frac{k_o t}{\phi \mu_o c_o \left(\frac{1}{c_o}\frac{\partial S_o}{\partial p} - S_o \right)} \qquad (3\text{-}31)$$

per unit radius squared.

The subsidiary equation is now introduced in this formula to account for the changing oil saturation within the interstices.

The variation of the diffusivity constant with the lowered reservoir pressure in Equation 3-30 will be treated in the nonlinear flow formulas.

Another consideration is the proportional constant and the physical parameters associated with the oil rate that applies in these equations. This refers to the $q_o\ \mu_o/4\pi\ kh$ term that the reader is familiar with in pressure buildup performance and to the basic treatment of the point-source solution by Lord Kelvin,[8] as well as the $Q/2\pi$ term described by the author.[6]

The point-source solution placed at the origin of a system, corresponding to Equation 3-30, is

$$\frac{Q}{2\pi} = \frac{\delta \rho_{oi} \delta x' \delta y'}{2\pi} \qquad (3\text{-}32)$$

Introducing the pore volume,

$$\frac{\phi h Q}{2\pi} = \frac{\delta \rho_{oi} \phi h \delta x' \delta y'}{2\pi} \tag{3-33}$$

For the oil density $\delta\, \rho_{oi} \;=\; -\; c_{oi}\; \rho_{oi}\; \delta p$, and

$$\frac{\phi h Q}{2\pi} = -\frac{c_{oi} \rho_{oi} \delta p \phi h \delta x' \delta y'}{2\pi} \tag{3-34}$$

Further,

$$\frac{1}{\rho_o} \frac{\partial (S_o \rho_o)}{\partial p} = c_o \left(\frac{1}{c_o} \frac{\partial S_o}{\partial p} - S_o \right) \tag{3-35}$$

and

$$\frac{Q c_{oi} \left(\frac{1}{c_o} \frac{\partial S_o}{\partial p} - S_o \right)_i \varphi h}{2\pi} = -c_{oi} \rho_{oi} \frac{\delta (S_o \rho_o)_i}{\rho_{oi}} \times \frac{\varphi h \delta x' \delta y'}{2\pi} \tag{3-36}$$

therefore,

$$-\frac{Q}{2\pi} = \frac{c_{oi} \rho_{oi} q_{oi} dt}{2\pi h \varphi c_{oi} \left(\frac{1}{c_o} \frac{\partial S_o}{\partial p} - S_o \right)_i} \tag{3-37}$$

where $$q_{oi} = \frac{1}{\rho_{oi}} \frac{\partial (S_o \rho_o)_i}{\partial t} \times \varphi h \delta x' \delta y'$$

is the rate of oil produced and voided at the point source.

Therefore, by the differentiation in Equation 3-30, relating absolute time with dimensionless time, we get

$$dt_D = \frac{k_o dt}{\varphi \mu_o c_{oi} \left(\frac{1}{c_o} \frac{\partial S_o}{\partial p} - S_o \right)_i} \tag{3-38}$$

and

$$-\frac{Q}{2\pi}=\frac{c_{oi}\rho_{oi}q_{oi}\mu_o dt_D}{2\pi k_o h} \tag{3-39}$$

The differential with respect to dimensionless time leads to the integration of the Ei function of the Lord Kelvin solution, leaving the coefficient $-c_{oi}\ \rho_{oi}\ q_{oi}\ \mu_o/2\pi\ k_o h$ as the rate relationship.

This is the comparable term developed earlier[6] that now includes density; but here an accounting is made for the oil saturation in place that is identified with the voidage that will occur in the reservoir.

The negative sign reverses the order of the density relationship in the nonlinear flow equations since the oil density increases as reservoir pressure declines (Fig. 3-4).

When all these factors are incorporated, including the variables for the nonlinear flow of gases, (see Chapter 2) then

$$\rho-\rho_j=\frac{c_{oj}\rho_{oi}q_{oi}\mu_{oj}}{2\pi k_{oj}h}\times\frac{1}{2}\times$$

$$\left[-Ei\left(\left(-\frac{\varphi\mu_o c_o}{k_o}\left(\frac{1}{c_o}\frac{\partial S_o}{\partial p}-S_o\right)\right)_j\frac{r^2}{4t}\right)+\right.$$

$$\left.Ei\left(\left(-\frac{\varphi\mu_o c_o}{k_o}\ \frac{1}{c_o}\frac{\partial S_o}{\partial p}-S_o\right)\right)_j\frac{r^2_j}{4t}\right)\right] \tag{3-40}$$

This represents the fluid movement in an infinite medium, subject to each increment of pressure drop in an areal traverse through the reservoir for a given time, with a constant rate of oil production.

This can be further simplified as expressed for pressure when we linearize the exponential relationship for the oil density in Equation 3-3.

Thus for the first pressure lowering from ρ_i to ρ_{j+1} using only the proportional terms involved in Equation 3-40.

$$(p_i-p_{j+1})\approx q_{oi} \tag{3-41}$$

For the second pressure lowering from p_{j+1} to p_j, then

$$(p_j - p_{j+1}) \approx c_{o(j+1)} \rho_{oi} q_{oi} \tag{3-42}$$

when expressed in densities.

Thus

$$(p_{j+1} - p_j) \approx \frac{\rho_{oi} q_{oi}}{\rho_{o(j+1)}} = q_{o(j+1)} \tag{3-43}$$

In Equation 3-43 it is accepted that the formation-volume factors for the oil are inversely proportional to the oil densities for the small changes in pressure encountered during the areal traverse of the reservoir.

For the pressure increment from p_j to p, again expressed in densities, then

$$(\rho - \rho_j) \approx c_j \rho_{o(j+1)} q_{o(j+1)} \tag{3-44}$$

This reflects the constancy in Q that whatever changes have occurred at higher pressures will prevail to lower pressures, unless intermediate changes occur to effect subsequent terms, and

$$(p_j - p) \approx \frac{\rho_{o(j+1)} q_{o(j+1)}}{\rho_{oj}} = q_{oj} \tag{3-45}$$

Therefore Equation 3-40, expressed in pressure, is the relationship

$$p_j - p = \frac{q_{oj}\mu_{oj}}{2\pi k_{oj} h} \times \frac{1}{2} \times$$

$$\left[-Ei\left(\left(-\frac{\varphi \mu_o c_o}{k_o} \left(\frac{1}{c_o} \frac{\partial S_o}{\partial p} - S_o \right) \right)_j \frac{r^2}{4t} \right) + \right.$$

$$\left. Ei\left(\left(-\frac{\varphi \mu_o c_o}{k_o} \left(\frac{1}{c_o} \frac{\partial S_o}{\partial p} - S_o \right) \right)_j \frac{r_j^2}{4t} \right) \right] \tag{3-46}$$

This is the form used to solve two-phase fluid flow discussed in the numerical example to follow.

Numerical example. The results for Field A are shown in Fig. 3-8 for transient fluid flow, representing two-phase fluids—the oil and its liberated solution gas flowing in the reservoir toward the well.

Production and reservoir data for the field are given in Table 3-3. These are shown as two listings, one in f-p-s units, familiar to the reader, and the other in c-g-s units, required to conform with the definition for permeability in darcys.

The production rate assigned in the problem is 50 b/d, producing

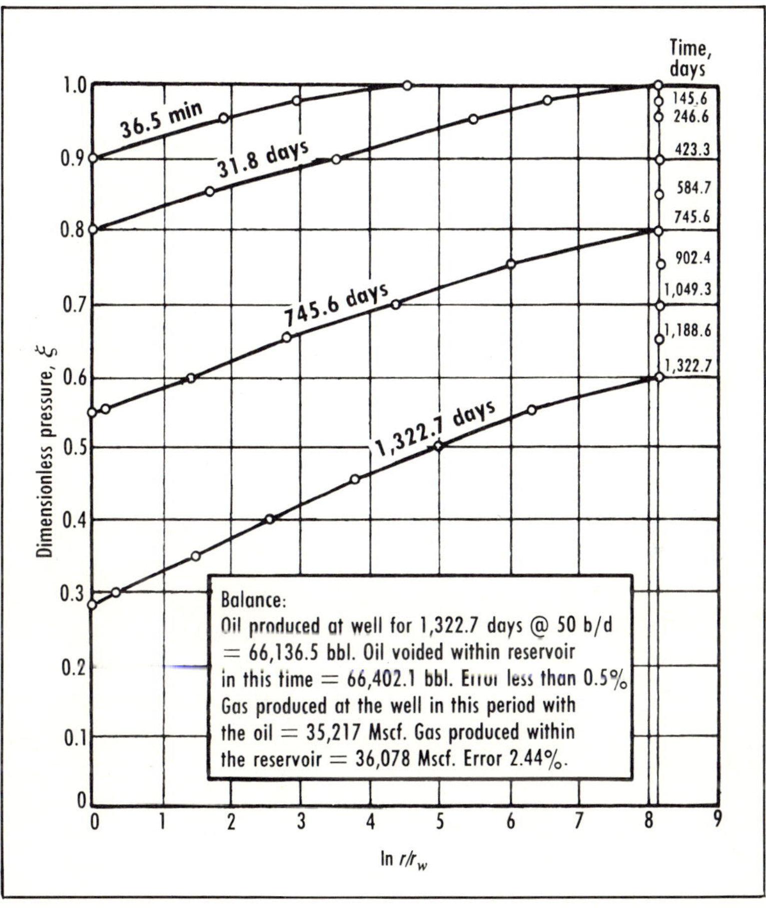

Fig. 3-8 Illustrative Example Of Oil and Its Solution Gas Flowing In A Reservoir.

TABLE 3-3
Production and reservoir data, Field A

f-p-s units		c-g-s units
q_o = 50 b/d	;	92.007 cc/sec
k_o = 173 md	;	0.173 darcys
h = 22.3 ft	;	679.704 cm
ϕ = 19.5%	;	0.195, fraction
S_w = 25%		0.25, fraction
r_w = 2.75 in.	;	6.985 cm
r_e = 776.83298 ft	;	23,677.86929 cm

from a net sand thickness of 22.3 ft, with 5½-in. casing set at top of the sand.

Initial well spacing was 2½ acres or less, but a spacing of 55 acres —exterior radius of 776.8 ft—is used in this example.

The permeability is 173 md, with the initial pressure of 434 psia, and the gas in solution is 76 scf/bbl.

The problem is the same as in Chapter 2 for gases; namely the areal traverses are established by pressure increments to determine radii for the absolute times of flow. For a fixed radius, such as at the well or the exterior boundary, these pressure increments are the independent variables that determine the times of pressure lowerings at these boundaries.

First, consider dimensionless time, t_D, given in Equation 3-31. To review briefly Figs. 3-1 to 3-3 show, by virtue of the subsidiary equation, a relationship between oil saturation and reservoir pressure.

Thus, by knowing oil saturation, the relative permeability for the oil is established (Fig. 3-6) that is related to pressure. See Table A-3, Appendix, for detailed calculations. Therefore each term in Equation 3-31 can be established from the listings in Table 3-2 to yield the coefficient associated with the absolute time in the equation.

These are instantaneous effects. In the application to nonlinear flow for small pressure changes, it is expedient to take average values for the beginning and end of each increment.

This likewise applies to the k_{oj}/μ_{oj}, in the rate relationship shown in Equation 3-46, as its instantaneous values from Table 3-2 and Fig. 3-6 can be averaged for the ensuing pressure change.

With respect to Fig. 3-8 and transient fluid flow, the initial calculations start at the well bore, r_w. Here the traverse is for a fixed radius using the $P(t_D)$ function of an earlier paper[9] that applies for the constant-rate case of fluids producing from an infinite medium in a radial system.

At the choice of the author this continued for a pressure lowering to $\xi = 0.80$, when the effects of an enclosed reservoir came into existence.

A time of 31.8 days and $\xi = 0.80$ (Fig. 3-8) is the basis for establishing the drainage radius, r_e. This follows from the drainage formula (Chapter 2) with the coefficient for t_D applying for the first pressure lowering from $\xi = 1.00$ to 0.975, to yield a drainage radius of 23,677.9 cm, or 776.8 ft. This is the enclosure for the illustrated problem.

In encountering the enclosure, both for its pressure lowering at this boundary with time and its areal sweep to the well bore, the mathematics used are the transcendental functions for a limited reservoir with the well producing at a constant rate.[9 10]

Next is the areal sweep for the infinite case represented by Equation 3-46.

The earlier treatments for the Ei functions are straightforward, but in the current problem we see a preliminary check on the work. In approaching the well bore for its fixed pressure $\xi = 0.80$, the calculated $r_w = 6.873$ cm, or 2.71-in. radius. The reported value is 2.75-in. radius, Table 3-4.

This likewise applies for the areal sweep for the time of 36.5 min. This curve is developed by the transformation of the curve for 31.8 days (Fig. 3-8). It was stated in Chapter 2 that for the infinite case one curve can be determined from the other for fixed ξ's by the relation that the radii squared are proportional to the times. As before, the radius realized in approaching the well for $\xi = 0.90$, is 2.71 in.

Then the gas produced at the well bore—for a constant rate of oil production—is related not only to the lowered pressures established in Fig. 3-8, but the physical data that define the gas/oil ratio at the well.

This gas/oil ratio is expressed as

$$GOR = \frac{k_g \mu_o B_o}{k_o \mu_g B_g} + R_s \qquad (3\text{-}47)$$

and Fig. 3-7 shows its variation with time that reflects the pressure changes and the gas produced at the well.

The insert in this plot gives the total amount of gas produced for the final time shown in Fig. 3-8; namely, 35,216,888 cu ft of gas in 1,322.7 days.

The calculations for Equation 3-47 depend upon the PVT data listed in Table 3-4. Also shown are the formation volume factors for the flash liberation that will be used in the material-balance application that will follow. These are established from the correlations given in Equations 3-26 and 3-27.

The intermediate pressures at the well bore for their corresponding times to determine gas/oil ratios have been interpolated by the Lagrangian Interpolations[11] using the data posted at the exterior radius and the calculated pressures at the well shown in Fig. 3-8.

Thus every inflection for gas/oil ratios illustrated in Fig. 3-7 are incorporated.

Balance. The accuracy of the results obtained in Fig. 3-8 for two-phase transient fluid flow must now be confirmed by the balance

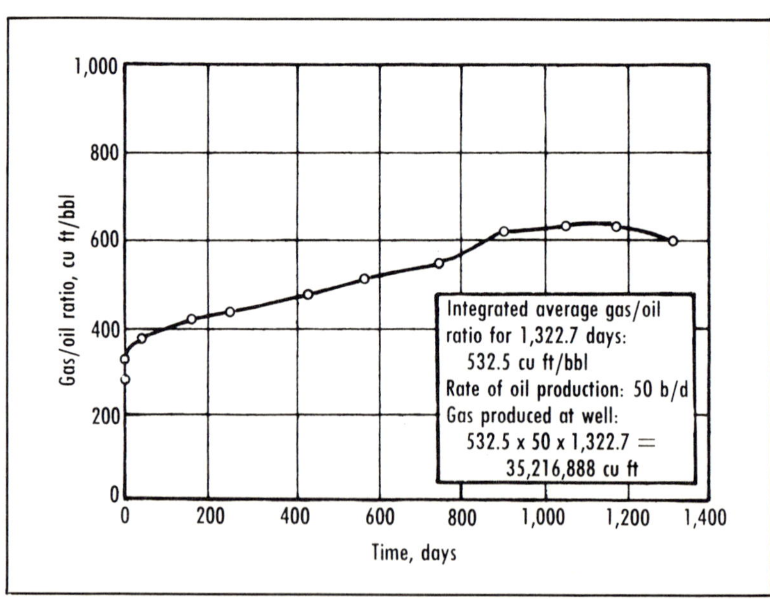

Fig. 3-7 Gas-oil Ratio Versus Time, Field.

between the oil and gas produced at the well with the voidage of the oil and gas from the reservoir pore space. This refers to the last pressure gradient shown in Fig. 3-8 for 1,322.7 days.

The first part of the balance relates to the oil voidage within the interstices, and as such must be represented as a dynamic condition described by the diffusivity equation, Equation 3-11, as well as the derivation for the point-source solution in defining the rate of oil produced.

This expression for oil voidage for a given radius and the lowering of pressure from the initial conditions, is given by

$$X_p = \int_{p_i}^{p} \frac{1}{B_o p_o} \frac{\partial(S_o p_o)}{\partial p} dp \qquad (3\text{-}48)$$

or

$$X_p = \int_{p_i}^{p} \frac{1}{B_o} \left(\frac{\partial S_o}{\partial p} - c_o S_o \right) dp \qquad (3\text{-}49)$$

where the factors related to this integrand are given in Tables 3-2 and 3-4. The sign is negative as would apply for the direction of pressure lowering and voidage but will be omitted.

Applied to the reservoir as a whole and the sweep along the final pressure gradient shown in Fig. 3-8, the formula is

$$N_p = \frac{\varphi h \times 2\pi}{159{,}000} \int_{0}^{In(r_e/r_w)} r^2 X_p d[In(r/r_w)] \qquad (3\text{-}50)$$

where

$$\frac{\varphi h \times 2\pi}{159{,}000} = \frac{(0.195)(679.704)(2\pi)}{159{,}000} = 523.76707(10)^{-5} \qquad (3\text{-}51)$$

These factors are listed in Table 3-3, and 1 bbl equals 159,000 cc.

TABLE 3-4
PVT data for Field A

ξ	B_o, Vol. @ res. cond. / Vol. @ std. cond.	B_g, Vol. @ res. cond. / Vol. @ std. cond.	R_s, Vol. / Vol. @ std. cond.	B_t, Vol. @ res. cond. / Vol. @ std. cond.
1.000	1.03500	0.031022	13.53613	1.03500
0.975	1.03459	0.032004	13.32069	1.04529
0.950	1.03409	0.033034	13.09408	1.05642
0.900	1.03298	0.035176	12.60294	1.08112
0.850	1.03178	0.037529	12.06920	1.10974
0.800	1.03054	0.040174	11.49719	1.14307
0.750	1.02924	0.043060	10.87881	1.18214
0.700	1.02790	0.046483	10.26044	1.22827
0.650	1.02653	0.050188	9.64207	1.28317
0.600	1.02514	0.054761	9.02369	1.34918
0.550	1.02372	0.059852	8.42089	1.42948
0.500	1.02227	0.066215	7.80689	1.52853
0.450	1.02081	0.073835	7.16686	1.65278
0.400	1.01932	0.082743	6.50036	1.81197
0.350	1.01782	0.096387	5.76680	2.02141
0.300	1.01655	0.11477	4.92365	2.30664
0.250	1.01477	0.13788	4.07339	2.71373
0.200	1.01322	0.17838	3.28701	3.33487
0.150	1.01166	0.24319	2.44215	4.38528
0.100	1.01008	0.39690	1.47290	6.51091
0.050	1.00849	0.84332	0.37931	12.94178
0.033871	1.00779	1.02885	0	19.08462

Therefore,

$$N_p = 523.76707(10)^{-5} \int_0^{In(r_e/r_w)} r^2 X_p d[(ln\ r/r_w)] \qquad (3\text{-}52)$$

and the results shown in the insert of Fig. 3-8, give 66,402.1 bbl of oil voided within the reservoir against 66,136.5 bbl of oil produced at the well in 1,322.7 days. This is an error of less than 0.5%.

To observe the gas produced from within the reservoir, the Schilthuis[12] material-balance equation has been applied.

This is a static condition illustrated by its author, where the pressure in the formation is everywhere uniform, and decreases in this manner with the fluids produced, comparable to the lowering of liquid in an enclosed tank.

This situation is not indicated in Fig. 3-8 except as this pressure gradient can equalize in infinite time.

This can be computed from the volumetric equation for the oil produced in 1,322.7 days to define the oil saturation in place, and the latter in turn is related to the uniform pressure that can exist throughout the reservoir.

Thus,

$$N_p = \frac{\varphi h r_e^2 \pi}{5.6146}\left(\frac{S_{oi}}{B_{oi}} - \frac{S_o}{B_o}\right) \tag{3-53}$$

and

$$66{,}136.4 = \frac{(0.195)(22.3)(776.83298)^2 \pi}{5.6146}\left(\frac{S_{oi}}{B_{oi}} - \frac{S_o}{B_o}\right) \tag{3-54}$$

where

$$\frac{S_{oi}}{B_{oi}} - \frac{S_o}{B_o} = 0.045042$$

Referring to Tables 3-2 and 3-4, $\xi = 0.54476$, or $p = 236.4$ psia that would prevail as a uniform pressure if the field were shut in at 1,322.7 days.

This is the basis for the material-balance equation following the concepts of Schilthuis, and it is expressed as

$$N(B_t - B_{ti}) = N_p[B_t - (R_p - R_{si})B_g] \tag{3-55}$$

The oil in place N can be computed from the data given in Tables 3-3 and 3-4, to be 1,064,023 bbl of oil corrected to standard conditions.

The interest here is to find R_p, the cumulative gas/oil ratio produced in 1,322.7 days.

For the computed average pressure that would apply in the Schilthuis formula, the oil produced in this time N_p, and the information listed in Table 3-4 for the physical factors, the cumulative gas/oil ratio calculated is 545.50997 scf/bbl.

This multiplied by 66,136.4 bbl of oil produced, gives 36,078,066 cu ft of gas.

The reference to Fig. 3-7 that is independent of the material-balance equation, but established for two-phase fluid flow and the changing pressures in the well, gives 35,216,888 cu ft of gas produced over this interval, or an error of 2.44% in the calculations for the gas produced.

What was shown. What has now been shown for two-phase fluid flow can apply to interference between wells in a reservoir and to include the configuration of a field. This is the image problem discussed in the previous chapter.

What has also been mentioned is that the variations in net sand thicknesses and permeabilities that occur within a reservoir can be accounted for in the mathematics.

In both these publications, Chapter 2 dealing with gases, and this presentation for two-phase fluid flow, a pattern is evidenced that indicates its adaptability to other problems in reservoir engineering through explicit formulas that apply for nonlinear flow and obey the boundary conditions.

NOMENCLATURE

q —rate of production, L^3/t
ρ —density, m/L^3
p —pressure, m/Lt^2
ξ —ratio of lowered to initial pressure, dimensionless
k —permeability, L^2
h —net thickness, L
ϕ —porosity, fraction
μ —viscosity, m/Lt
c —compressibility, Lt^2/m
S_o —oil saturation, fraction
S_w —connate water, fraction
r —radius, L
r_w —well radius, L
r_e —radius of external boundary, L
t —absolute time, t
t_D —dimensionless time, based on unit radius
B —formation volume factor
R —gas-oil ratio

SUBSCRIPTS

i —initial condition
j —subsequent condition
o —oil
g —gas
s —solution gas
t —flash liberation

REFERENCES

1. Sir Horace Lamb, Hydrodynamics, New York Dover Publications, 1945.

2. William Hurst, "Unsteady Flow of Fluids in Oil Reservoirs," Physics Vol. 5, January 1934.

3. John L. Kennedy, "Here's the way reserves for the big new Hilight Field were estimated," OGJ, May 18, 1970.

4. James W. Amyx, Daniel M. Bass, Jr., and Robert L. Whiting, Petroleum Reservoir Engineering, McGraw-Hill Book Co. Inc., N.Y., 1960.

5. E. T. Whittaker and G. Robinson, The Calculus of Observations, Blackie & Son Ltd., England, 1944.

6. William Hurst, "Interference Between Oil Fields," Trans. AIME, Vol. 219, 1960.

7. William Hurst, "The Solution of Non-Linear Equations," SPE 3676, Presented Sept. 13, 1971, Anadarko Basin Section of the AIME, Liberal, Kan.

8. H. S. Carslaw, Introduction to the Mathematical Theory of the Conduction of Heat in Solids, MacMillian & Co. Ltd., London, 1921.

9. A. F. van Everdingen and W. Hurst, "The Application of the Laplace Transformations to Flow Problems in Reservoirs," Trans. AIME, Vol. 186, 1949.

10. Morris Muskat, "The Flow of Compressible Fluids Through Porous Media and Some Problems in Heat Conduction," Physics, Vol. 5, March 1934.

11. Tables of Lagrangian Interpolation Coefficients, National Bureau of Standards, Columbia Univ. Press, N.Y., 1944.

12. Ralph J. Schilthuis, "Active Oil and Reservoir Energy," Trans. AIME, Vols. 118 to 123, 1936-1937 (Reprint).

4

IN-FILL DRILLING

IN EVERY OIL FIELD, depleted or producing, with water drive or not, oil is left behind. This is oil dormant in a field, that has never entered into the mainstream of fluid flow to the producing wells, and even though pressure depletion occurs and gas is weathered off, the oil does not move.

This chapter points out how such areas can exist through the development of streamlines showing oil movement to producing wells within the field proper.

The scope of this work is based on conformal mapping[1] presented in Chapter 1. The potential problem makes the determination of streamline development a practical application.

ILLUSTRATIVE PROBLEM

The illustrative problem under discussion is shown in Figs. 4-1 and 4-2. This is the off-center well already referred to in Chapter 1 where the well location is $(r, \theta) = (50, 45°)$ in an enclosed radius of $r_e = 100$ for a circle. Its corresponding location on the conformal rectangle is likewise shown in Fig. 4-2, with the line AB the reproduction of the areal extent of the same line in Fig. 4-1. Points A, B, 1, 2, 3, and the Well are identical points in both domains. With reference to these points, calculations performed for the rectangle give numerical values that check with Muskat's solution for an off-center well in the enclosed circle. This is another sustainment for conformal mapping.

An illustrative problem will now be presented showing the streamlines as they extend out from the producing well to the confines of the enclosures for the field. Fig. 4-3 refers to the rectangle upon

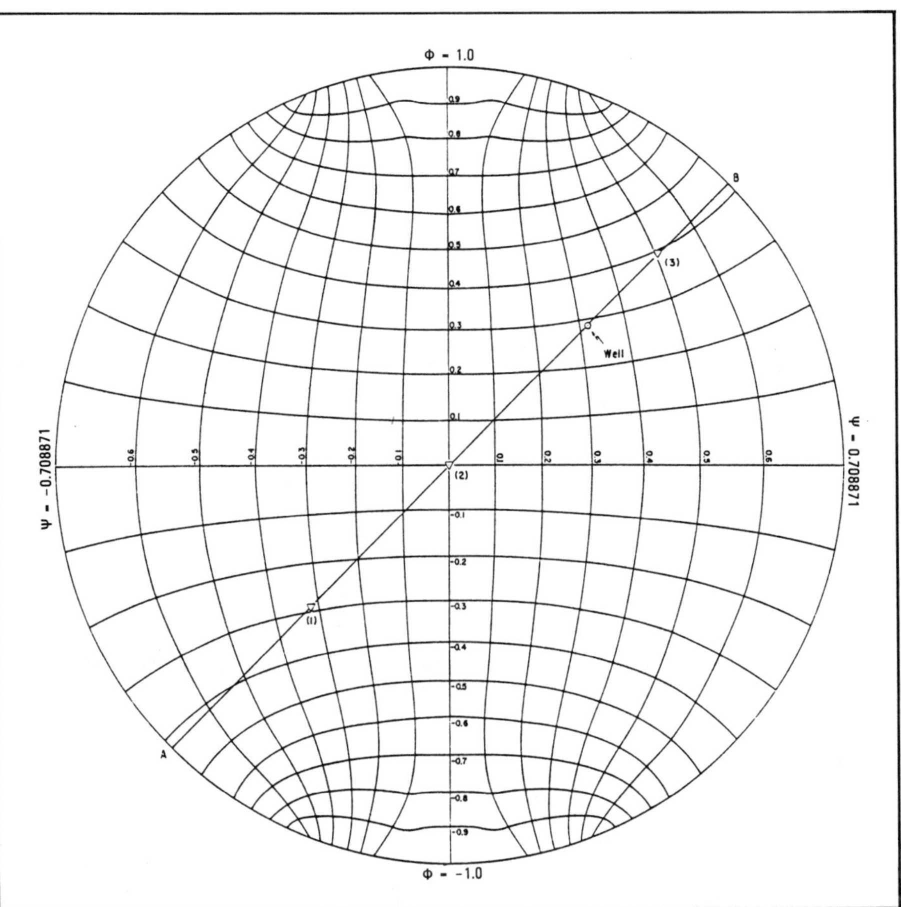

Fig. 4-1 Illustration of potential and streamline distributions within a circle. The line source and sink are located on the circumference opposite one another, and each subtends 45°.

which the basic work was performed, and Fig. 4-4 its conformal transference to the circle as represented by Fig. 4-1.

The uniformity of Fig. 4-3 is impressive. Note that radial lines extend outward from the center of the rectangle, and these revolve in a counterclockwise direction like a radar scope. The calculations are performed on equidistant points to determine the resultant of fluid flow within the rectangle. Where the resultant is the same as the

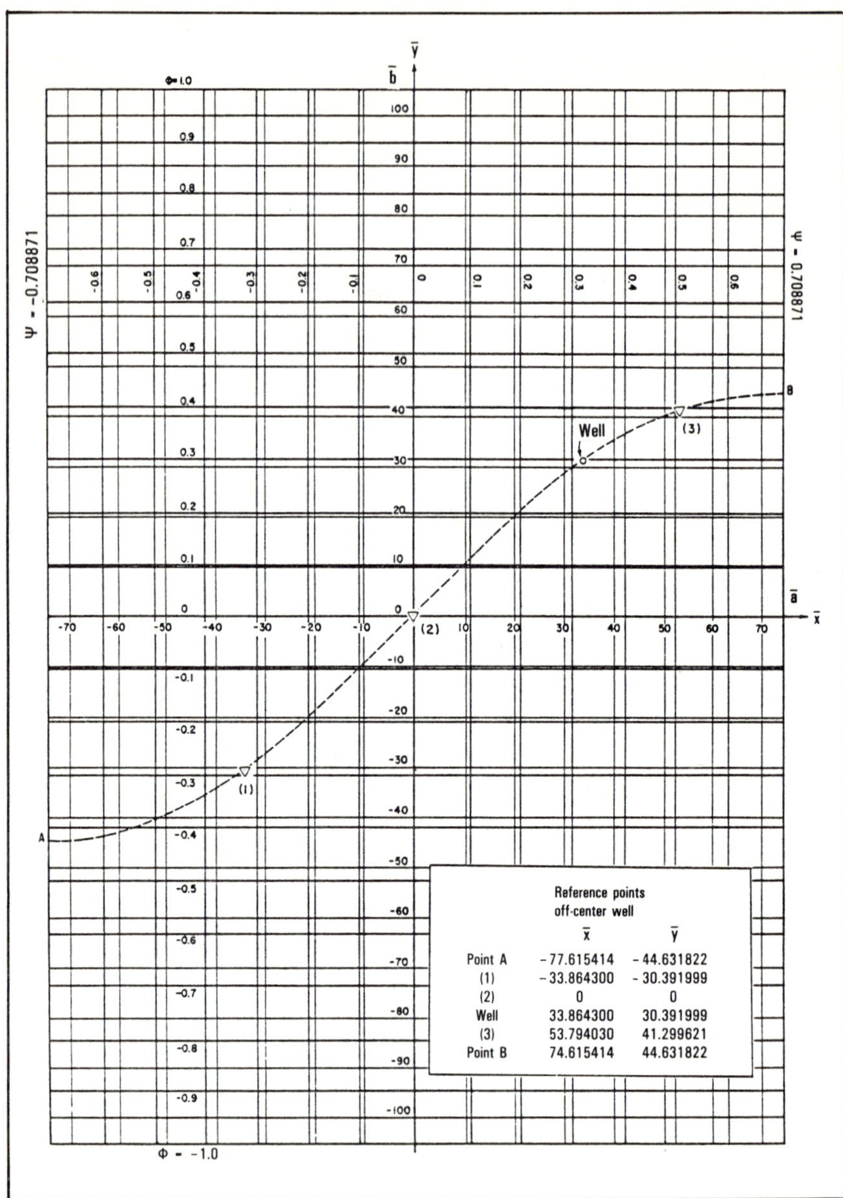

Fig. 4-2 Conformal rectangle reproduced for the circle, $r_e = 100$.

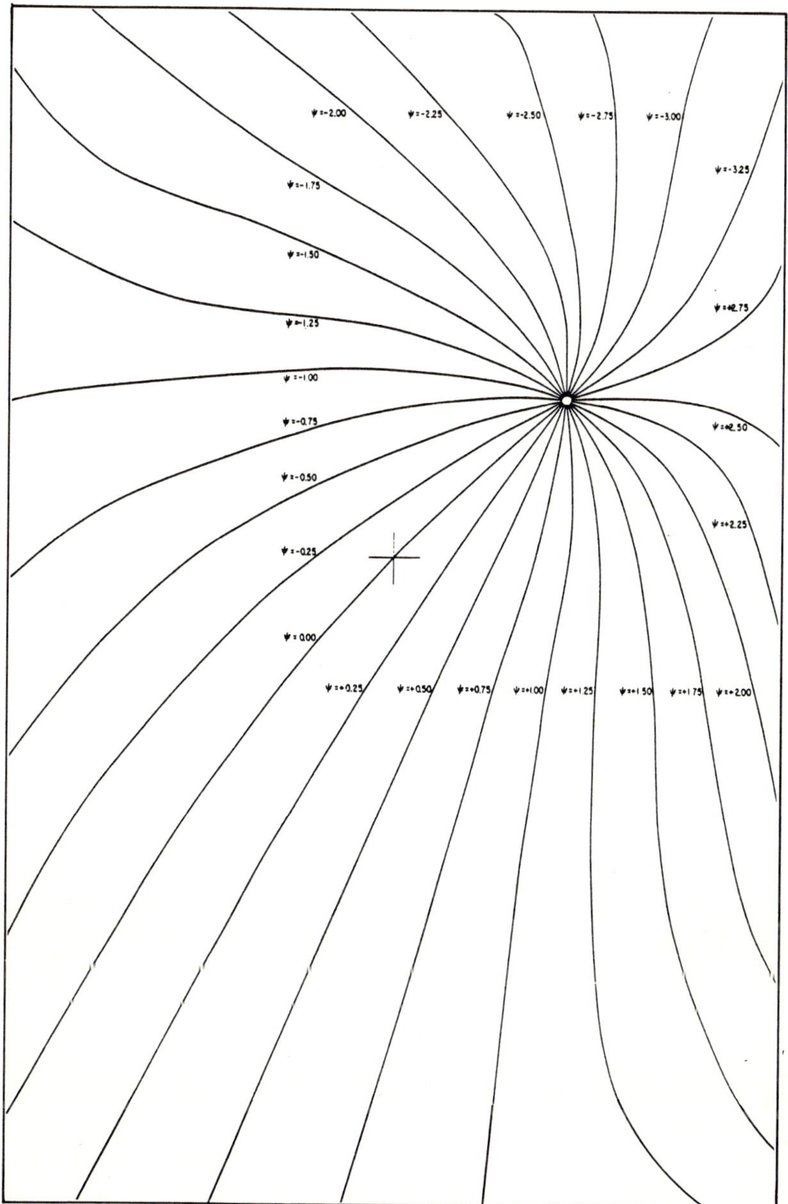

Fig. 4-3 Illustration of streamlines for off-center well.

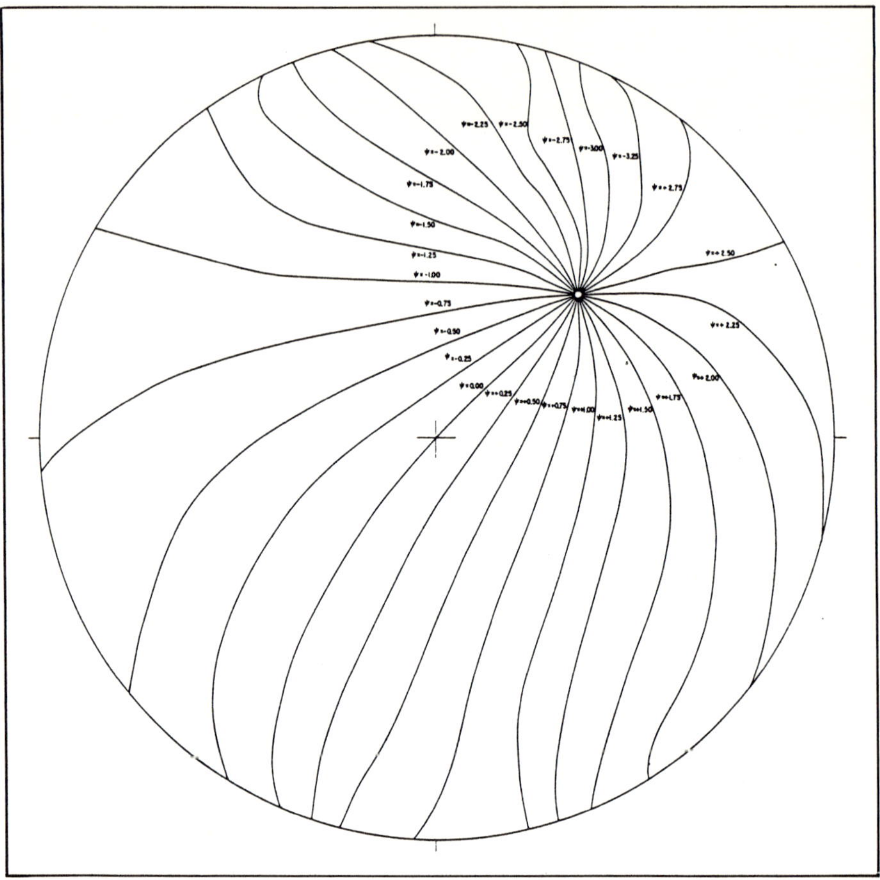

Fig. 4-4 Streamlines from rectangle reproduced in the field.

direction of the moving scope, these streamlines are positive; in the opposite direction, they are negative, explaining the difference in signs shown on the plot.

Table 4-1 summarizes the results obtained in this plot. These refer to the fluid flux at each of the extended coordinates at 90° taken in a positive direction from 0° around the rectangle. These fluid fluxes are expressed in radians, and the calculations show that for the first quadrant this difference is $3\pi/2$, and for each of the succeeding quadrants the difference is $\pi/2$.

TABLE 4-1
Calculated flux on rectangle

Degrees	Points, $(\bar{x}, \bar{y})$	Calculated flux, ψ
0	74.615414, 0	+2.350078
90	0, 105.25951	−2.357858
180	−74.615414, 0	−0.783946
270	0, −105.25951	+0.783768

The first quadrant shows the total fluid fluxes from the entry of all three quadrants, with itself supplying the last entry. The total flux around the well itself is 2π. The reader may observe the cardinal values reported on the plot is the difference between $\Psi = 2.75$ and -3.25 or $\Psi = 6.00$ which approaches 2π.

The transfer of the streamlines shown in Fig. 4-3 to Fig. 4-4, which is the field for the off-center well in the enclosed circle, is given by

$$\frac{\phi}{\bar{y}} = \frac{1}{105.25951} = 0.00950033 \tag{4-1}$$

$$\frac{\psi}{\bar{x}} = \frac{0.708871}{74.615414} = 0.00950033 \tag{4-2}$$

These equations identify the Φ, Ψ, values in the circle for the $\bar{x}$, $\bar{y}$ coordinates in the conformal rectangle to reproduce the streamlines for the field.

The uniformity for the streamlines shown in Fig. 4-3 is no longer present in Fig. 4-4. Here we observe wide gaps in areas for the consecutive values for fluid flux that are prime candidates for in-fill drilling. Such are recognized for streamlines $\Psi = 2.25$ to 2.75 and particularly for streamlines $\Psi = -0.50$ to -1.25.

Considering the closeness of the well to the boundary of the circle, it is not surprising this occurs, since the well itself is subject to the wide expansion of the field's production. This is comparable to an early publication[2] by van Everdingen and the writer showing that in gas cycling the Sheridan Field, Colorado County, Texas, the loop on streamlines upstream from the gas in-put well never seems to close off; wet gas could still be produced even through cycling was at an end.

A comparable analysis is the East Texas field where 3,000 wells are abandoned. Operators could no longer afford to produce oil at high water cuts for $3/bbl. Pockets of oil can exist, however, behind every producing well up-stream from the water drive in the Woodbine. For the entrepreneurs, leasing is too unattractive to exploit this oil, as royalty interests are too diversified. This is a problem for the oil companies who still maintain property rights, and the Federal Government who can expropriate land to exploit this oil.

What is described for the East Texas field can apply to any field where water drive has occurred; oil can exist upstream from the water drive behind a producing well.

For the opposite direction in the region of $\Psi = -0.50$ and -1.25, in Fig. 4-4, the wide gaps in areal extent are developed from the nature of the problem, for which the writer offers no explanation but is part of the problem and the solution.

ANALYTICS

The analytics follow from conformal mapping. This is expressed by the $P(t_D)$ function as

$$P(r,t_D) = a + bt_D - \Delta\phi \tag{4-3}$$

where

$$\Delta\phi = \phi(\bar{x}, \bar{y}) - \phi(\bar{x}_w, \bar{y}_w) \tag{4-4}$$

and

$$\phi(\bar{x},\bar{y}) = \frac{1}{2} \sum_{j=-2}^{j=+2} \ln\left\{\cosh\frac{\pi\bar{y}}{2\bar{a}} - \cos\frac{\pi}{2\bar{a}}(\bar{x}-\xi)\right\} + \ln\left\{\cosh\frac{\pi\bar{y}}{2\bar{a}} + \cos\frac{\pi}{2\bar{a}}(\bar{x}+\xi)\right\} \tag{4-5}$$

the application of Basset's formula.

The first term in Equation 4-3 is the semisteady state in the linear depletion of the well in the field, and the second term by this potential function Equation 4-4 is the transference of this semistate depletion to any portion of the field $(\bar{x}, \bar{y})$ to observe pressure drop from the subject well.

By application of methods in Chapter 1, the entire range of transient fluid flow can be reproduced at a point from the semistate pressure line, by superimposing the difference between the actual well decline of Fig. 1-7 and this semistate line to observe the entire range of transient fluid flow from the inception of pressure decline at that point.

Fig. 4-5 illustrates the introduction of the Basset Formula. Point sources of the well are placed above and below the well in addition to the field itself. These are rapidly convergent series, such that few tiers are used, two above the field, the field itself, and two tiers below. This concept is employed such that no fluid will flow across the boundary of the field along the extremities of the $\bar{x}$ axis.

The value $\phi(\bar{x}_w, \bar{y}_w)$ is fixed for the well and represents the distances to the well radius location on the cartesian coordinates. It conforms to Equation 4-5.

Table 4-2 shows the comparison of the $P(t_D)$ term for the identified points on Figs. 4-1 and 4-2, starting with the rectangle, and how these compare with Muskat's[3] solution on the circle. The comparison is excellent. These correspond to $t_D = 15{,}000$, the transit-time value for which Fig. 1-12 was plotted (Chapter 1).

The streamlines as herein presented are the essential interest of this chapter. The differential of Equation 4-3, since $\bar{x}$ and $\bar{y}$ in that equation are the only variables, is given as

$$\frac{d\phi}{d\bar{x}} = \frac{1}{2} \times \frac{\pi}{2\bar{a}} \sum_{j=-2}^{j=+2} \left\{ \frac{\sin \frac{\pi}{2\bar{a}}(\bar{x}-\xi)}{\cosh \frac{\pi\bar{y}}{2\bar{a}} - \cos \frac{\pi}{2\bar{a}}(\bar{x}-\xi)} - \frac{\sin \frac{\pi}{2a}(\bar{x}-\xi)}{\cosh \frac{\pi\bar{y}}{2\bar{a}} + \cos \frac{\pi}{2\bar{a}}(\bar{x}-\xi)} \right\} \tag{4-6}$$

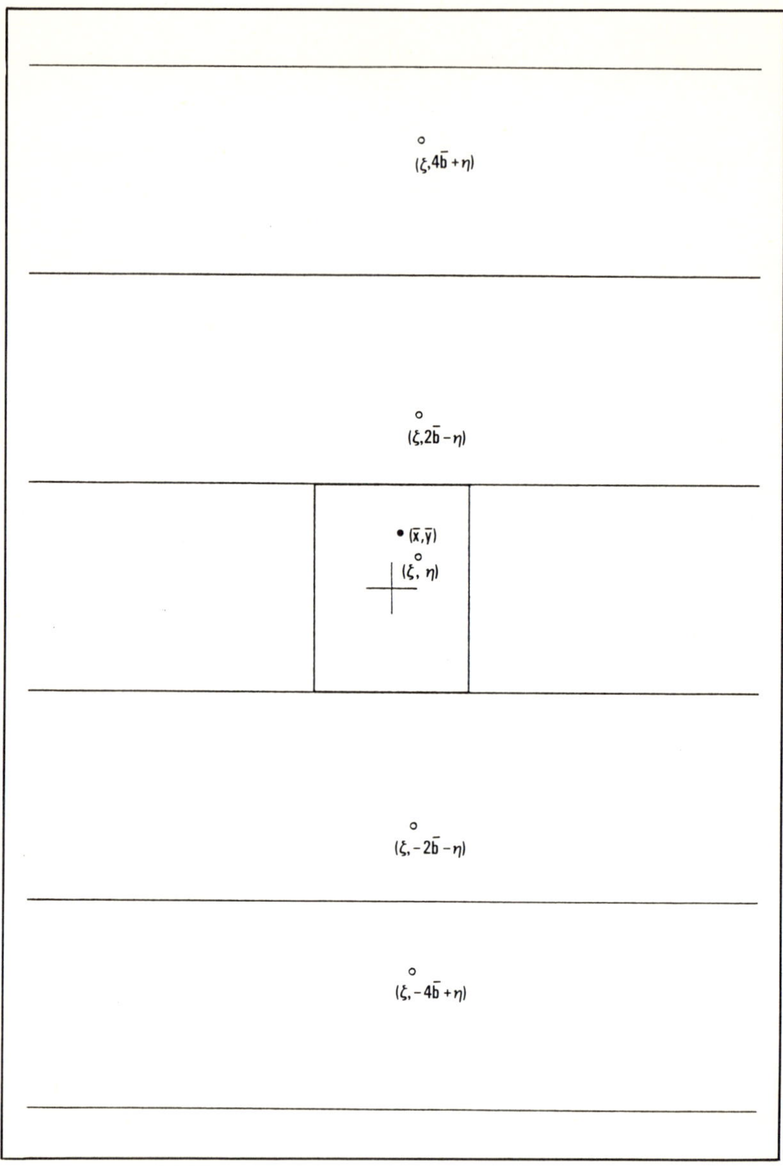

Fig. 4-5 Location of the point sources with respect to the enclosed rectangle.

TABLE 4-2

Comparison of the calculations performed by Conformal mapping with Muskat's solution for an off-center well

Reference Points	$(\bar{r}, \bar{\theta})$, Fig. 4-1	$(\bar{x}, \bar{y})$, Fig. 4-2		$P(t_D)$ Muskat's Solution	$P(t_D)$ Conformal Mapping Solution	Rate of fluid flow on AB, $d\phi/dr$ Muskat's Solution	Rate of fluid flow on AB, $d\phi/dr$ Conformal Mapping Solution
Point A	100,225°	−74.615414	−44.631822	2.064070	2.039938	0	0
(1)	50,225	−33.864300	−30.391999	2.276856	2.380106	−0.00884625	−0.00987429
(2)	0,225	0	0	3.068147	3.286215	−0.0250	−0.0254669
Well	50,45	33.864300	30.391999	7.404591	7.404591	1.00	1.00
(3)	75,45	53.794030	41.299621	4.512548	4.527719	+0.0253571	+0.0252828
Point B	100,45	74.615414	44.631822	4.261294	4.172058	0	0

where the algebraic sign determines the direction of fluid infringement on the radar scope. The ξ coordinate is fixed, but the $\bar{y}$ coordinates for the point sources must account for the distances from η_j to $\bar{y}$, or $(\eta_j - \bar{y})$ to the point $(\bar{x}, \bar{y})$, as this is fluid flow between bounded planes. The η_j's as defined for η_1, η_2, and η_{-1}, η_{-2} are $2\bar{b} - \eta$, $4_{\bar{b}} + \eta$, $-2\bar{b} - \eta$, and $-4\bar{b} + \eta$ in conformance with Fig. 4-5.

The velocity in the $\bar{y}$ direction is the same identity as Fig. 4-5, as the point sources are now oriented with respect to the $\bar{y}$ axis, such that no fluid flow will take place along the enclosures in that direction. The equation is

$$\frac{\overline{d\phi}}{d\bar{y}} = -\frac{1}{2} \times \frac{\pi}{2\bar{b}} \sum_{k=-2}^{k=+2} \left\{ \frac{\sin \frac{\pi(\bar{y}-\eta)}{2\bar{b}}}{\cosh \frac{\pi x}{2\bar{b}} - \cos \frac{\pi(\bar{y}+\eta)}{2\bar{b}}} - \frac{\sin \frac{\pi(\bar{y}+\eta)}{2\bar{b}}}{\cosh \frac{\pi x}{2\bar{b}} + \cos \frac{\pi(\bar{y}+\eta)}{2\bar{b}}} \right\} \tag{4-7}$$

The sign in this equation has been changed to identify the fluid flux as positive with the direction of the rotation of the scope.

The resultant for both of these equations is expressed as

$$\frac{d\phi}{d\bar{r}} = \sqrt{\left(\frac{d\phi}{d\bar{x}}\right)^2 + \left(\frac{d\phi}{d\bar{y}}\right)^2} \tag{4-8}$$

and the normal to the rotating arm is

$$\frac{d\phi}{dN} = \frac{d\phi}{dr} \sin(\tau - \theta) \tag{4-9}$$

where

$$\tau = \tan^{-1} \frac{\frac{d\phi}{d\bar{y}}}{\frac{d\phi}{d\bar{x}}} \tag{4-10}$$

and ϕ is the angle of the direction of the rotator as shown in Fig 4-6.

The total fluid flux is given by

$$\psi = \int_{o}^{\bar{r}} \frac{d\phi}{dN} \times d\bar{r} \qquad (4\text{-}11)$$

which explains the equal distances plotted on the rotating arm to establish this integration by Simpson's rule.

These are the reported results shown in Figs. 4-3 and 4-4, expressed in radians.

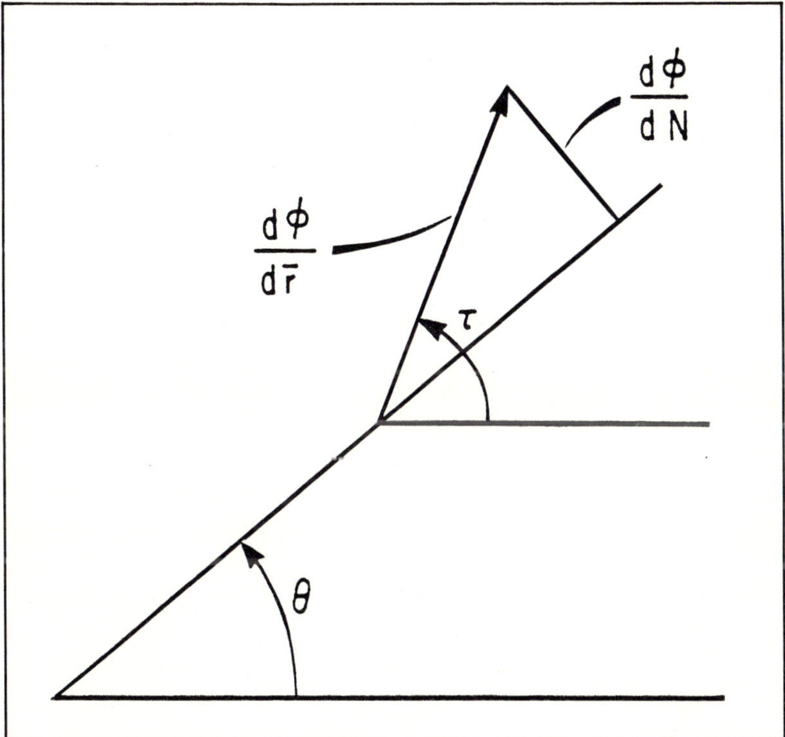

Fig. 4-6 The fluid rate normal to the rotating arm.

We will now determine the velocity of fluid flow from the rectangle to Muskat's[3] solution on the circle. These are the reference points already referred to.

For the coordinates shown in Fig. 4-2, the resultant velocity can be calculated from Equations 4-6, 4-7 and 4-8 for its values along the curvature AB in Fig. 4-2. Equation 10 establishes τ and the angle θ between the straight line and a potential ϕ in Fig. 4-1 is read by a protractor for each of these representative points, as angles in conformal mapping do not change. What is represented in Fig. 4-1 is the same for Fig. 4-2, that applies for the curvature AB on the rectangle.

This velocity potential projected on AB is

$$\frac{d\phi}{d\bar{r}} \cos(\tau - \theta)$$

There is one other item to be considered to establish the comparison with Muskat's solution and this calculation on the rectangle, and that is the linear distances in each of these domains and the interchange of r for $\bar{r}$.

This is the conformal representation of

$$\frac{d\bar{r}}{dr} = \frac{1}{0.00950033} \sqrt{\left(\frac{d\phi}{dr}\right)^2 + \left(\frac{d\psi}{dr}\right)^2} \tag{4-12}$$

and applied as the chain rule gives the identical velocities as shown in Table 4-2 for Muskat's solution for velocities in a circle for an off-center well.

It should be of interest to the reader to know that in this entire analysis, formation capacity, kh, permeability times sand thickness, cancels out. This applies whether we are dealing with absolute permeability or relative permeability and for all variations in net sand thicknesses.

The criterion is the difference rates as established for each well in the field with respect to its orientation from the boundaries, and this includes faults and shale lines that exist in all fields and can be handled by conformal mapping. It should be mentioned that faults and shale lines accentuate the spread between consecutive streamlines because the fluid must flow around these barriers to reach the wells and these are likely areas for in-fill drilling.

To explain this, the coefficient associated with Equation 4-3 that applies to Equations 4-5 and 4-6 is $qu/2\pi kh$. In applying Darcy's law we have

$$u = -\frac{kh}{\mu}\frac{d\psi}{dx} \tag{4-13}$$

The multiplication of this fluid rate by the potential velocity yields $q/2\pi$ associated with the flow formula; namely, the actual well production rate divided by 2π.

Thus in each of these plots shown for Figs. 4-3 and 4-4, to establish the flow rate across any curvature or lease line, the difference between $\Delta\psi$ shown on these plots is multiplied by the reference well q, and divided by 2π. This gives the actual oil rate across any designated line or curvature in these domains. This is expressed as

$$\text{Oil Rate} = \frac{q}{2\pi}\Delta\psi \tag{4-14}$$

The reference to the actual well rate is an arbitrary assigned nomenclature for a chosen well, as every well in the field is brought into this complex by its reference to this unit well. This is to expedite computer application in not having to deal with large numbers, as our interest is to expose wide areas where in fill well drilling can be developed.

Computers are applicable because we are treating with explicit formulas that are programmable. Our objective is to deal with any number of wells in any field by using terminals for programming. Our interest is not to produce massive print-outs that cause confusion rather than interest, but results that expedite plotting of these streamlines.

It should be mentioned that the streamlines coming into the producing well, Figs. 4-3 and 4-4, are not graphical interpretations but calculations of angles of influx, whether for the rectangle or the circle. Each streamline can be identified with its entry into a well bore, and as such its range of fluid flow to the well from the field's enclosures is selective. It is not as shown for the illustrative problem to cover a field-wide drainage, but selective areas of drainage as develop in the field for each well in this complex of wells.

The Texas Railroad Commission and other regulatory bodies have extensive files on depleted oil fields, showing their geology and production. These data are immediately applicable, using the well's production rates in their heyday, to the search for dormant oil that has never entered into the mainstream of fluid flow in a field. As illustrated in this paper this is the basis for in-fill drilling.

In reviewing the entire annals of reservoir engineering, this is the first application of this art where oil can be found.

REFERENCES

1. Conformal Mapping of Oil and Gas Fields, by William Hurst, presented before the Annual Convention of the Society of Petroleum Evaluation Engineers, Hilton Inn, Dallas, Texas, April 25, 1977.

2. Performance of Distillate Reservoirs in Gas Cycling, by W. Hurst, and A. F. van Everdingen, Trans. A.I.M.E., 1946.

3. M. Muskat, Flow of Homogeneous Fluids, Text, J. W. Edwards, Inc., 1946.

APPENDIX

DEVELOPMENT OF MATHEMATICAL FORMULAS, DATA, AND NUMERICAL PROBLEMS

DIFFUSIVITY EQUATION

Fig. A-1 represents an infinitesimal element of a reservoir. A fluid balance on this element describes total fluid flow and the voidage in the element itself.

The fluid flow rate is expressed as mass velocity, where ρ is the density of the fluid, and u and v are the velocities along the OX and OY axis, is given as ρU, or ρV along the respective axis.

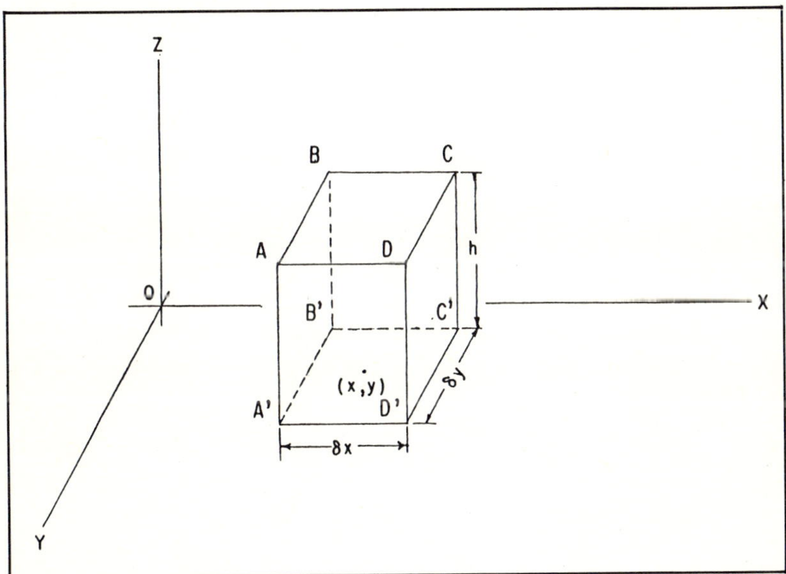

Fig. A-1 The diffusivity equation.

Thus, the fluid flow outward from the face DCC′D′ is given as

$$\left[pu + \frac{1}{2}\frac{\partial(\rho u)}{\partial x}\delta x\right] h\delta y \tag{A-1}$$

For fluid flowing inward to the element along face ABB′A′

$$\left[pu - \frac{1}{2}\frac{\partial(\rho u)}{\partial x}\delta x\right] h\delta y \tag{A-2}$$

The total loss of fluid in this direction is the difference between Equations A-1 and A-2, or

$$\frac{\partial(\rho u)}{\partial x} h\delta\times\delta y$$

This procedure applied to the OY direction yields

$$\frac{\partial(\rho v)}{\partial y} h\delta\times\delta y$$

The sum of these two values is the voidage from the element, and

$$\left[\frac{\partial(\rho u)}{\partial x} + \frac{\partial(\rho v)}{\partial y}\right] h\delta\times\delta y = \frac{-\partial\rho}{\delta t} h\delta\times\delta y \tag{A-3}$$

or

$$\frac{\partial(\rho u)}{\partial x} + \frac{\partial(\rho v)}{\partial y} = \frac{-\partial\rho}{\partial t}$$

The forms that the reader is familiar with in this equation are

$$\frac{\partial^2 p}{\partial x^2} + \frac{\partial^2 p}{\partial y^2} = \frac{\partial p}{\partial t_D} \tag{A-4}$$

where

$$u = -\frac{k}{\mu}\frac{\partial p}{\partial x}; \qquad v = -\frac{k}{\mu}\frac{\partial p}{\partial y}$$

Darcy's law, and

$$t_D = \frac{kt}{\phi\mu\,(1 - S_w)c} \tag{A-5}$$

for dimensionless time, if we specify unit radius in the denominator of Equation A-5.

likewise, where density is expressed as an exponential form

$$p = p_{ie} - c(pi - p)$$

where c is fluid compressibility, then Equation A-4 takes the form

$$\frac{\partial^2 \rho}{\partial x^2} + \frac{\partial^2 \rho}{\partial y^2} = \frac{\partial \rho}{\partial t_D} \quad \text{(A-6)}$$

Its transformation into radial or circular coordinates is given by

$$\frac{\partial^2 \rho}{\partial r^2} + \frac{1}{r}\frac{\partial \rho}{\partial r} = \frac{\partial \rho}{\partial t_D} \quad \text{(A-7)}$$

with which the reader is familiar.

LORD KELVIN'S SOLUTION

In the middle part of the nineteenth century, Lord Kelvin undertook the solution of Equation A-4 as applied to heat.

It is supposed that he accepted a solution to Equation A-4 as

$$p = A(t_D)e - B(t_D)(x^2 + y^2) \quad \text{(A-8)}$$

where $A(t_D)$ & $B(t_D)$ are functions of dimensionless time.

The substitution into Equation A-4 yields

$$\frac{\partial p}{\partial x} = A(t_D)B(t_D)e^{-B(t_D)(x^2+y^2)}(-2x)$$

and

$$\frac{\partial^2 p}{\partial x^2} = A(t_D)B^2(t_D)e^{-B(t_D)(x^2+y^2)}(4x^2)$$
$$+ A(t_D)B(t_D)e^{-B(t_D)(x^2+y^2)}(-2)$$

The same could be applied to $\partial^2 p/\partial y^2$

With respect to time, this follows as

$$\frac{\partial p}{\partial t_D} = A^1(t_D)e^{-B(t_D)(x^2+y^2)}$$
$$+A(t_D)B^1(t_D)e^{-B(t_D)\,(x^2+y^2)}(-(x^2+y^2))$$

When these terms are collected in Equation A-4, then

$$A(t_D)B^2(t_D)4(x^2+y^2)+A(t_D)B(t_D)(-4)$$
$$=A^1(t_D)+A(t_D)B^1(t_D)(-(x^2+y^2))$$

Therefore, recognizing what are the variables, the following relations are separated

$$4A(t_D)B^2(t_D) = -A(t_D)B^1(t_D) \qquad \text{(A-9)}$$

and

$$A(t_D)B(t_D)(-4) = A^1(t_D) \qquad \text{(A-10)}$$

It follows from Equation A-9 that

$$\frac{\partial B(t_D)}{\partial t_D} = -4B^2(t_D)$$

and

$$B(t_D) - \frac{1}{4t_D}$$

From this relation and Equation A-10

$$\frac{1}{A(t_D)}\frac{\partial A(t_D)}{\partial t_D} = -\frac{1}{t_D}$$

and

$$In\ A(t_D) = -In\ t_D$$

or,

$$A(t_D) = \frac{1}{t_D}$$

Replacing these terms in Equation A-8, then Lord Kelvin's equation takes the form

$$p = \frac{1}{t_D}e^{-\frac{(x^2+y^2)}{4t_D}} \qquad \text{(A-11)}$$

We must also include the dissipation of fluid at the source into the formation proper, as shown in Fig. A-2.

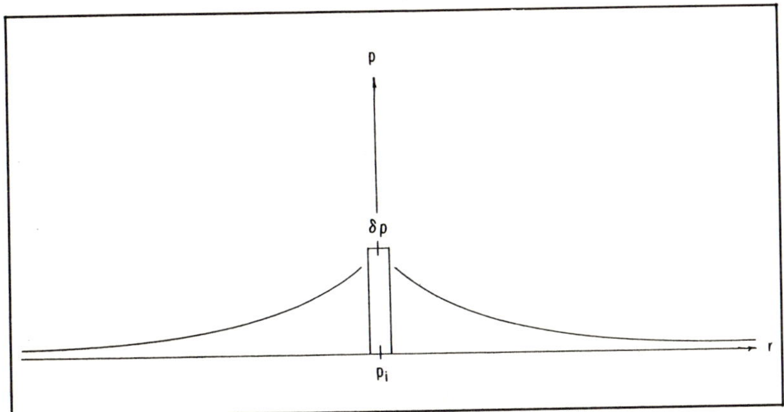

Fig. A-2 Dispersment of fluid in the reservoir.

This is given as

$$p = \frac{\delta p \bar{h}^{2}}{C t_D} e^{-\frac{(x^2+y^2)}{4t_D}} \tag{A-12}$$

where $\partial p \bar{h}^{2}$ is the fluid placed at the origin which will totally dissipate into the formation, with C a constant to be established. That is

$$\delta p \bar{h}^{2} = \int_0^\infty 2\pi \; prdr = \int_0^\infty 2\pi \frac{\delta p \bar{h}^{2}}{C t_D} e - \frac{r^2}{4t_D} rdr \tag{A-13}$$

if we let $u = \frac{r^2}{4t_D}$, and $du = \frac{r}{2t_D} dr$, this equation becomes

$$1 = \frac{4\pi}{} \int_0^\infty e^{-u} du = \frac{4\pi}{C} \left[-e^{-u} \right] = \frac{4\pi}{C}$$

and C = 4π. Thus, Equation A-12 takes the form of

$$p = \frac{\delta p \bar{h}^{2}}{4\pi t_D} e^{-\frac{(x^2+y^2)}{4t_D}} \tag{A-14}$$

exactly as is given in Carslaw, the Conduction of Heat, p. 150, MacMillian and Co., Ltd., 1921.

This is still in the abstract. What Equation A-14 shows is an increment of pressure change ∂p, applied to a small cross section area of $\bar{h}^2$ at the origin.

If we convert this to a column of fluid of height, h, and porosity ϕ, then

$$p = \frac{\delta p \bar{h}^{2} \phi h}{4\pi\phi h t_D} e^{-\frac{(x^2+y^2)}{4t_D}} \qquad \text{(A-15)}$$

Therefore, this increment of pressure δp associated with the column of fluid has meaning. We will associate this with the equation of state for fluid compressibility; namely,

$$v = v_o\,[1 + c(pi - p)]$$

Since we will always be dealing with cumulative pressure drop from an initial pressure, p_i, then

$$\delta v = v_o c \delta \Delta p$$

Therefore, Equation A-15 transposed gives

$$\Delta p = \frac{\delta v}{4\pi\phi c h t_D} e^{-\frac{(x^2+y^2)}{4t_D}} \qquad \text{(A-16)}$$

Therefore, this δv is a volume of infinitesimal fluid by virtue of the change in ∂p pumped into the reservoir at the origin.

These are initial-pressure conditions, so δv is at p_i.

What this means, is that

$$\delta v = q_i dt$$

is a constant rate of fluid injected initially. Translated in Equation A-16 and changing absolute time t, to dimensionless time, t_D, then

$$\Delta p = \frac{q_i \mu}{4\pi k h t_D} e^{-\frac{(x^2+y^2)}{4t_D}} dt_D$$

If this is done continuously, injecting fluid into the origin with time, then

$$\Delta p = \frac{q_i \mu}{4\pi kh} \int_0^{t_D} \frac{e^{-\frac{(x^2+y^2)}{4(t_D - t_D')}}}{(t_D - t_D')} dt'_D$$

Making the necessary calculus interchanges, then

$$\Delta p = \frac{q_i \mu}{4\pi kh} \int_{\frac{r^2}{4t_D}}^{\infty} \frac{e^{-\eta}}{\eta} dn \qquad \text{(A-17)}$$

The logarithmic integral which is the Ei function, or

$$\Delta p = \frac{q_i \mu}{2\pi kn} \times \frac{1}{2} \left\{ -Ei \left(\frac{-r^2}{4t_D} \right) \right\} \qquad \text{(A-18)}$$

The $P(t_D)$ function that we are dealing with in this book is expressed by

$$P(r,t_D) = \frac{1}{2} \left\{ -Ei \left(\frac{-r^2}{4t_D} \right) \right\} \qquad \text{(A-19)}$$

THE SOLUTION OF − Ei (− z)

The following discussion refers to Equations 1-5 and 1-6 in Chapter 1 on conformal mapping. Also see Table A-1.

The integral − Ei (− z) can be broken up into sub-integrals; thus

$$-Ei(-z) = \int_z^{\infty} \frac{e^{-z}}{z} dz$$

$$= \int_1^{\infty} \frac{e^{-z} dz}{z} - \int_0^1 (1 - e^{-z}) \frac{dz}{z} - \int_1^z \frac{dz}{z} \qquad \text{(A-20)}$$

$$+ \int_0^z (1 - e^{-z}) \frac{dz}{z}$$

or collecting parts we observe that for theintegrand $\frac{dz}{z}$, we have

$$-\int_0^1 - \int_1^z + \int_0^z = -\int_0^1 + \int_0^1 = 0$$

and for the term dz is is the summation

$$\int_1^\infty + \int_0^1 - \int_0^z = \int_z^\infty \frac{e^{-z}}{z}\,dz$$

the integral we wish to determine.

The reason for breaking the integral in this manner is that the first two terms in Equation A-20 constitute Euler's Constant, where

$$\lim_{v \to Inf.} \left(1 + \frac{1}{2} + \frac{1}{3} + \frac{1}{4} + \ldots \frac{1}{v} - Inv\right) = \gamma = 0.577216$$

shown in Chapter 1.

This is the starting point; thus

$$1 + \frac{1}{2} + \frac{1}{3} + \frac{1}{4} \cdots + \frac{1}{v} = \int_0^z (1 + x + x^2 + \ldots + x^{v-1})dx \qquad \text{(A-21)}$$

$$= \int_0^1 \frac{(1 - x^v)}{1 - x}\,dx$$

Introducing into Equation A-21

$$x = 1 - \frac{z}{v}$$

the term on the right becomes

$$\int_0^v \frac{[1-(1-\frac{z}{v})^v]}{z}\,dz = \int_0^1 \frac{[1-(1-\frac{z}{v})^v]}{z}\,dz + \int_1^v \frac{[1-(1-\frac{z}{v})^v]}{z}\,dz$$

If v goes to infinity as shown for Euler's Constant, then the above equation becomes

$$=\int_0^1 \frac{(1-e^{-z})}{z}\,dz + \ln v - \int_1^v \frac{e^{-z}}{z}\,dz$$

where

$$\lim_{v \to Inf.} \left(1-\frac{z}{v}\right)^v = e^{-z}$$

and

$$\ln v = \int_1^v \frac{dz}{z}$$

Coming back to Equation A-20, we see that the first two terms in that formula, Euler's Constant, are negative.

Thus, we can account for this and the remaining terms therein shown, where

$$-Ei(-z) = -\gamma - \ln z + \int_0^z (1-e^{-z})\frac{dz}{z}$$

This last integral is simply applied.

$$\int_0^z (1-e^{-z})\frac{dz}{z} = \int_0^z \left[1-1+\frac{z}{1!}-\frac{z^2}{2!}+\frac{z^3}{3!}\right]\frac{dz}{z}$$

$$= \int_0^z \left[\frac{1}{1!}-\frac{z}{2!}+\frac{z^2}{3!}+\frac{z^3}{4!}\right]dz$$

$$= z - \frac{z^2}{2 \cdot 2!} + \frac{z^3}{3 \cdot 3!} - \frac{z^4}{4 \cdot 4!}$$

$$= -\sum_{n=1}^{\infty} (-1)^n \frac{z^n}{n \cdot n!}$$

Therefore, this whole procedure can be summarized, to give

$$-Ei(-z) = -\gamma - \ln z - \sum_{n=1}^{Inf.} (-1)^n \frac{z^n}{n \cdot n!} \qquad \text{(A-22)}$$

as given in Equation 1-5 in Chapter 1.

The assymptotic expansion, Equation 1-6, is simply solved. Given

$$y=F(z)=e^{+z}\int_z^\infty \frac{e^{-z}}{z}\,dz$$

$$y^1=F^1(z)=e^{+z}\int_z^\infty \frac{e^{z}}{z}\,dz-\frac{1}{z}$$

$$\therefore y^1=y-\frac{1}{z}$$

or,

$$y^1-y+\frac{1}{z}=0 \qquad \text{(A-23)}$$

Let

$$y=a_o+\frac{a_1}{z}+\frac{a_2}{z^2}+\frac{a_3}{z^3}$$

$$y^1=-\frac{a_1}{z^2}-\frac{2a_2}{z^3}-\frac{3a_3}{z^4}$$

Therefore, by Equation A-23, we have

$$-\frac{a_1}{z^2}-\frac{2a_2}{z^3}-\frac{3a_3}{z^4}-a_o-\frac{a_1}{z}-\frac{a_2}{z^2}-\frac{a_3}{z^3}+\frac{1}{z}=0$$

and

$$\begin{aligned} a_0 &= 0 \\ a_1 &= 1 \\ a_2 &= -a_1 \\ a_3 &= -2a_2 \\ a_4 &= -3a_3 \end{aligned}$$

$$\therefore F(z)=\frac{1}{z}-\frac{1}{z^2}+\frac{2!}{z^3}-\frac{2\cdot 3}{z^4}$$

and

$$\int_z^\infty \frac{e^{-z}}{z}\,dz=e^{-2}\left[\frac{1}{z}-\frac{1}{z^2}+\frac{2!}{z^3}-\frac{3!}{z^4}+\ldots\right] \qquad \text{(A-24)}$$

THE BASSET FORMULA

For steady-state flow for a well located between fixed boundaries, defined by Basset, and illustrated in Fig. A-3, the relationship is

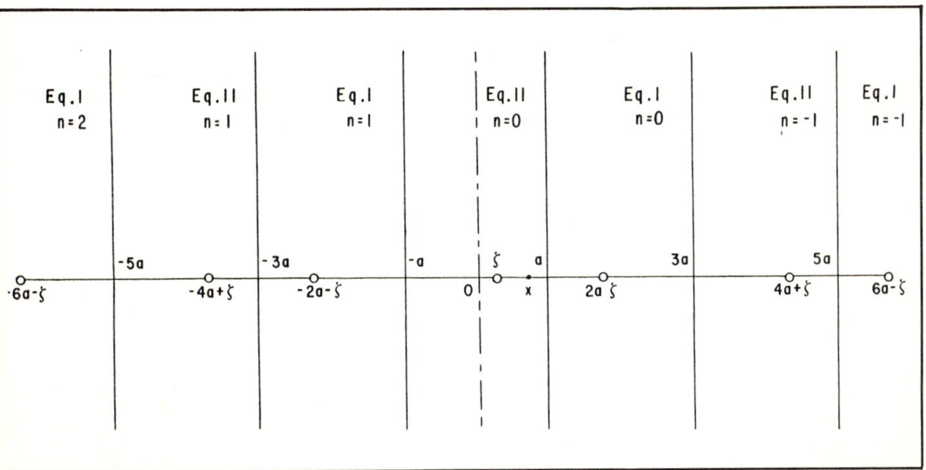

Fig. A-3 Basset formula.

$$\phi = \frac{1}{2} \sum_{-\infty}^{+\infty} In \left\{(x+\zeta-2a+4na)^2+y^2\right\} + \frac{1}{2} \sum_{-\infty}^{+\infty} In \left\{(x-\zeta+4na)^2+y^2\right\} \qquad \text{(A-25)}$$

If we express these arguments within the logarithmic terms as apply to the x-axis, given by

$$\lambda = (x+\zeta-2a+4na) \qquad \text{(I)}$$

$$\eta = (x-\zeta+4na) \qquad \text{(II)}$$

the reader will observe that these summations in Equation A-25 obey all the locations of the image wells for $n = -\infty$ to $n = +\infty$, including $n = 0$. Application of Equations I and II is shown in Fig. A-3.

Therefore, Equation A-25 can be expressed as a product; namely

$$\phi = \frac{1}{2} In \prod_{-\infty}^{+\infty} \left\{(x-\zeta+4na)^2+y^2\right\} + \frac{1}{2} In \prod_{-\infty}^{+\infty} \left\{x+\zeta-2a+4na)^2+y^2\right\}$$

interchanging the order in which these series will be taken.

Thus,

$$\phi = \frac{1}{2} In \prod_{+\infty}^{\infty} \left\{ \left(\frac{(x-\zeta)}{4na} + 1 \right)^2 + \frac{y^2}{4n^2a^2} \right\} + \frac{1}{2} In \prod_{+\infty}^{\infty} (4n^2a^2)$$

$$+ \frac{1}{2} In \prod_{-\infty}^{\infty} \left\{ \left(\frac{x+\zeta-2a}{4na} + 1 \right)^2 + \frac{y^2}{4n^2a^2} \right\} + \frac{1}{2} In \prod_{-\infty}^{\infty} (4^2n^2a^2)$$

The constants $\Pi(4^2n^2a)$ will be excluded, as they serve no part in getting the potential differences in Fig. A-3.

Therefore,

$$\phi = \frac{1}{2} In \prod_{-\infty}^{\infty} \left\{ \left(\frac{x-\zeta}{4na} + 1 + \frac{iy}{4na} \right) \left(\frac{x-\zeta}{4na} + 1 - \frac{iy}{4na} \right) \right\}$$

$$+ \frac{1}{2} In \prod_{-\infty}^{\infty} \left\{ \left(\frac{x+\zeta-2a}{4na} + 1 + iy \right) \left(\frac{x+\zeta-2a}{4na} + 1 - \frac{iy}{4na} \right) \right\}$$

$$= \frac{1}{2} In \prod_{-\infty}^{\infty} \left\{ \left(\frac{(x-\zeta+iy)}{4na} + 1 \right) \left(\frac{(x+\zeta-iy)}{4na} + 1 \right) \right\}$$

$$+ \frac{1}{2} In \prod_{-\infty}^{\infty} \left\{ \left(\frac{(x+\zeta-2a+iy)}{4na} + 1 \right) \left(\frac{(x+\zeta-2a-iy)}{4na} + 1 \right) \right\} \quad \text{(A-26)}$$

This relationship in Equation A-26 is the product of the series expansion given by Basset, or

$$sin \frac{\Pi\theta}{c} = \frac{\Pi\theta}{c} \prod_{-\infty}^{\infty}{}' \left(1 + \frac{\Theta}{nc} \right) \quad \text{(A-27)}$$

where Π' denotes n = o is excluded.

Therefore, Equation A-26 is again reproduced as

$$\phi = \frac{1}{2}\, ln \frac{(x-\zeta+iy)}{4a} \prod_{-\infty}^{\infty}{}' \left(\frac{(x-\zeta+iy)}{4na}+1\right)$$

$$+\frac{1}{2}\, ln \frac{(x-\zeta-iy)}{4a} \prod_{-\infty}^{\infty}{}' \left(\frac{(x-\zeta-iy)}{4na}+1\right)$$

$$+\frac{1}{2}\, ln \left(\frac{(x+\zeta-2a)+iy}{4a}\right) \prod_{-\infty}^{\infty}{}' \left(\frac{(x+\zeta-2a+iy)}{4na}+1\right)$$

$$+\frac{1}{2}\, ln \left(\frac{(x+\zeta-2a-iy)}{4a}\right) \prod_{-\infty}^{\infty}{}' \left(\frac{(x+\zeta-2a-iy)}{4na}+1\right)$$

Therefore, the series expansion of Equation A-27 comes into effect; or

$$\phi = \frac{1}{2}\, ln\left(sin\, \pi\, \frac{(x-\zeta+iy)}{4a}\, sin\, \pi\, \frac{(x-\zeta-iy)}{4a}\right)$$
$$+\frac{1}{2}\, ln\left(sin\, \pi\, \frac{(x+\zeta-2a+iy)}{4a}\, sin\, \pi\, \frac{(x+\zeta-2a-iy)}{4a}\right) \qquad \text{(A-28)}$$

Using the expression

$$Sin\, \pi\, \frac{(x-\zeta+iy)}{4a} = sin\, \pi\, \frac{(x-\zeta)}{4a}\, cosh\, \frac{\pi y}{4a} + icos\, \pi\, \frac{(x+\zeta)}{4a}\, sin\, h\, \frac{\pi y}{4a}$$

and

$$sin\, \pi\, \frac{(x-\zeta-iy)}{4a} = sin\, \pi\, \frac{(x-\zeta)}{4a}\, cosh\, \frac{\pi y}{4a} - i\, cos\, \pi\, \frac{(x-\zeta)}{4a}\, sin\, h\, \frac{\pi y}{4a}$$

available in any integral table, and cross-multiplying as indicated, then

$$cosh^2 \frac{\pi y}{4a} - cos^2 \frac{\pi(x-\zeta)}{4a} = cosh\, \frac{\pi y}{2a} - cos\, \frac{\pi(x-\zeta)}{2a}$$

This is for the first term in Equation A-28. Likewise, for the second term, we have

$$cosh\, \frac{\pi y}{2a} + cos\, \frac{\pi(x+\zeta)}{2a}$$

Expressing these in the logarithmic function, we obtain the Basset Formula, or

$$\phi = \frac{1}{2} In \left\{ cosh \frac{\pi y}{2a} - cos\ \pi \frac{(x-\zeta)}{2a} \right\} \qquad \text{(A-29)}$$

$$+ \frac{1}{2} In \left\{ cosh \frac{\pi y}{2a} + cos\ \pi \frac{(x+\zeta)}{2a} \right\}$$

Sample Problem
(See Fig. 3-8)

TABLE A-1
Numerical Calculation of $-Ei(-z)$ function

z	Eq. 1-5	Eq. 1-6	Actual $-Ei(-z)$	Terms in Series Expansion in Eq. 1-5
0.25	1.044283		1.044321	4
0.50	0.559725		0.559774	4
0.75	0.339983		0.340341	4
1.00	0.217923		0.219384	4
1.00	0.219393		0.219384	8
1.25	0.142091		0.146413	4
1.25	0.146436		0.146413	8
1.50	0.0895845		0.100020	4
1.50	0.100060		0.100020	8
1.75	0.0475893		0.0694888	4
1.75	0.0695433		0.0694888	8
2.00	0.0489317		0.0489005	8
2.50	0.0243704		0.0249149	8
3.00	0.0130478		0.0130484	15
3.50	0.00697107	0.00636397	0.00697014	15
4.00	0.00378950	0.00346995	0.00377935	15
	Maximum Limit: 5.50			
5.50	0.00224137	0.000630280	0.000640926	15

Remarks:

z = 5.50 is the final point for calculation. Intervening values for z

z	$-E_i(-z)$
4.50	0.00207340
5.00	0.10114830

TABLE A-2
Muskat's solution, off-center well in Fig. 1-12, Chapter 1

	r	θ	$P(t_D)$ $t_D = 15{,}000$
Point A	100	225°	2.064070
	90	225°	2.071964
	80	225°	2.096163
	70	225°	2.137574
	60	225°	2.197326
(1)	50	225°	2.276856
	40	225°	2.378039
	30	225°	2.503382
	20	225°	2.656365
	10	225°	2.842035
(2)	0	225°	3.068147
	5	45°	3.200076
	10	45°	3.347584
	15	45°	3.514034
	20	45°	3.704333
	25	45°	3.926076
	30	45°	4.191957
	35	45°	4.525742
	40	45°	4.980729
	45	45°	5.726875
Well	49	45°	7.381258
Well	51	45°	7.404591
	55	45°	5.843566
	60	45°	5.214260
	65	45°	4.876413
	70	45°	4.660221
(3)	75	45°	4.512548
	80	45°	4.409798
	85	45°	4.339457
	90	45°	4.294128
	95	45°	4.260115
Point B	100	45°	4.261294

TABLE A-3

Permeability and liquid saturation data used to plot Fig. 3-6, Chapter 3.
k = 173 md, from core analysis
Where: $C_{w} = 25\%$

ξ	S_o	$S_o + C_w$	Relative permeability, md
1.00	0.75000	1.00000	173
0.975	0.74578	0.99578	166
0.950	0.74207	0.99207	160
0.900	0.73545	0.98545	151
0.850	0.72928	0.97928	145
0.800	0.72319	0.97319	138
0.750	0.71728	0.96728	132
0.700	0.71171	0.96171	126
0.650	0.70636	0.95636	121
0.600	0.70120	0.95120	116
0.550	0.69611	0.94611	111
0.500	0.69124	0.94134	107
0.450	0.68664	0.93664	103
0.400	0.68176	0.93176	99
0.350	0.67645	0.92645	94
0.300	0.67061	0.92061	91
0.250	0.66426	0.91426	86
0.200	0.65758	0.90758	82
0.150	0.64999	0.89999	77
0.100	0.64081	0.89081	71
0.050	0.62775	0.87775	64
0.033871	0.62147	0.87147	61

Determination of $$\frac{1}{K_o} = \frac{\phi \mu_0 c_0}{k_0}\left(\frac{1}{c_0}\frac{\partial S_0}{\partial p} - S_0\right)$$

from Chapter 3. Converting pressure to atmospheres, the constant is $\phi \times 14.7 = 0.197 \times 14.7 = 2.8665$.

ζ	$\frac{1}{K_o} = \frac{\mu_0}{k_0}\left(\frac{\partial S_0}{\partial p} - S_0 c_0\right) \cdot 2.8665$
1.000	0.0372941
0.975	0.0315982
0.950	0.0296190
0.90	0.0268408
0.850	0.0275342
0.800	0.0295763

0.750	0.0290780
0.700	0.0288180
0.650	0.0289027
0.600	0.0294719
0.550	0.0301479
0.500	0.0289026
0.450	0.0301615
0.400	0.0351881
0.350	0.0405133
0.300	0.0475588
0.250	0.0541573
0.200	0.0617126
0.150	0.0797987
0.100	0.119486
0.050	0.213341
0.033871	0.281440

Then,

ξ	K_o	K_o (Average)
1.00	26.813893	29.230633
0.975	31.647372	32.704742
0.950	33.762112	35.509413
0.900	37.256714	36.787592
0.850	36.318469	35.064662
0.800	33.810855	34.10058
0.750	34.390261	34.545398
0.700	34.700534	34.649689
0.650	34.598844	34.264735
0.600	33.930625	33.550216
0.550	33.169806	33.884385
0.500	34.598963	33.876907
0.450	33.154850	30.786774
0.400	28.418698	26.550975
0.350	24.683252	22.854928
0.300	21.026603	19.745667
0.250	18.464731	17.334439
0.200	16.204146	14.367839
0.150	12.531532	10.450357
0.100	8.369181	6.528256
0.050	4.687332	4.120243
0.033871	3.553155	

Determination of

$$(p_j - p) \Big/ \frac{q_{oj}\mu_{oj}}{2\pi k_{oj} h}$$

for the illustrated example in Chapter 3,

ξ	$(p_i - p) \Big/ \frac{q_{oj}\mu_{oj}}{2\pi k_{oj} h}$
1.00 to .975	1.06116
0.975 to .950	1.01790
0.950 to .900	1.93389
0.900 to .850	1.82987
0.850 to .800	1.73874
0.800 to .750	1.64821
0.750 to .700	1.56451
0.700 to .650	1.48733
0.650 to .600	1.41686
0.600 to .550	1.34750
0.550 to .500	1.28508
0.500 to .450	1.22929
0.450 to .400	1.17815
0.400 to .350	1.12431
0.350 to .300	1.06427

These tables are used for classroom demonstration to determine distance of enclosure as well as areal transverse for the cited problem in Chapter III.

INDEX